Becoming

If we could read the thoughts of Aku, leader of a small clan of Sapiens living in the fertile valley of Mesopotamia on the shore of the Black Sea 7500 years ago, it might read something like..

Life is good. It is getting warmer and food is plentiful. For some time now members of my group have heard stories from our trading partners that the Sea Without End, beyond the great hill where the sun vanishes, is reclaiming the land. I indeed have verified this on my recent voyage. I have given much thought to this matter. Perhaps the Sea Without End God is at war with our Sea God. I must encourage all to pray to our God to give her strength and help her protect our valley and our way of life. But the Sea Without End God is very powerful and my clan must reconnect with our old ways and be prepared to move quickly and go to higher ground. If we need to move, I hope our God will give me a vision. It would be nice if the Gods would stop fighting with each other, but I guess that is better than them fighting us. They make life tough sometimes. Some days I am not sure I really believe in these gods; but then how else would I and all I see, be?

Even though the Black Sea God lost and the flood wiped out many other tribes, Aku's tribe did survive. Those genes of Aku's people became incorporated into me. So it is important to consider how Aku came to be, indeed how our universe came to be. How did anything happen?

Acknowledgements:

Thanks to members of the Palouse Writers League, who read and critiqued earlier versions of chapters in this book. Thanks also to members of the Moscow Coffee Club, Lance and Dominick, to Chuck Hower, and to my sons Tom and Pat for their comments on the draft version of the book. These comments were invaluable in producing this final piece. The patience and encouragement of my wife, Kathleen, during this process was amazing.

The Amazon self-publishing company, Create Space, has created tutorials and easy-to-follow step-by-step instructions to facilitate the entire publishing process. They made the process fun.

I relied on many sources for the information used to put together the stories in this book. None of the words used here are original but I hope the grouping of the words and the ideas are.

I am reminded of the worker who, after installing new tile on our kitchen floor, commented *I did the very best I could, I hope you like it*. That also is my hope for you the reader.

October, 2016   Bob Kearney

Becoming

Bob Kearney

	Page
Prologue	5
I. Becoming	7
Chapter 1: Overview	8
II. Our Universe	13
Chapter 2: Observations	14
Chapter 3: The Experiments	19
Chapter 4: The Creation Story	27
III. Life	47
Chapter 5: Life on Earth	48
Chapter 6: Bacteria and Archaea	52
Chapter 7: Viruses	57
Chapter 8: Recent Thoughts and Studies	61
Chapter 9: History of Life	71
Chapter 10: Killer Viruses	79
IV. Consciousness	87
Chapter 11: Consciousness	88
Chapter 12: The Human Journey	102
Chapter 13: Human Environments	113
Chapter 14: Brains and the Mind	123
Chapter 15: Myths and Religions	133
Chapter 16: The Future	157

References 168

Tables and Figures

Table 1: Time-Line of Major Events 103

Figure 1: Earth Temperature vs. Time 114

Prologue

The title of this story has changed from Beginning to Becoming and back again a number of times. It started as a story of how I, an 80-year-old man raised in an Irish Catholic family, became a non-practicing atheist. I soon realized this journey took about 70 years and that it was a gradual process, not a sudden realization. Oh sure, there are some vivid moments that I remember, like being an altar-boy for one week before my folks were told by the parish priest that I did not seem to be cut out for that and I should do something else. But there was certainly no ah-ha moment; no sudden realization that I needed to start a critical examination of my life and beliefs. I think I was too pleased to just be a kid, then too happy when I found my life's companion, Kathy, and too pleased with raising a family.

It was not until I reached my late 50s and could actually imagine retirement, that I started to have serious conversations with my mind, to investigate my cognitive map and question all my beliefs. Along with that I decided to make an effort to "walk in the shoes of others". Perhaps that was motivated by the joke, first told to me by my late brother-in-law John, "If you are going to continue to disagree with someone first you need to walk a mile in their shoes. If you still disagree, remember you are a mile away and the other person has no shoes." But, it soon was no joke for me and made me more aware of how my brain, my own cognitive map, structured my beliefs. I realized I could change that map if I wanted.

My four grandparents, born in the western part of Ireland in the late 1800s, immigrated to the US shortly before the 20th century started. There, two by two, they met, begat my parents who begat me. By my counting that makes me second generation American and at least fourth generation Irish. I still don't know why I cannot automatically claim Irish citizenship.

Please keep in mind the words of that Irish writer, Maeve Binchy, "*I had a very happy childhood, which is unsuitable if you're going to be an Irish writer.*" Like Binchy I had a happy childhood, full of memories like the moments when I discovered that ice cream came in flavors other than vanilla and then again when I concluded that I actually liked math. But those memories are for a different story.

Turn the page to start Becoming.

I. Becoming

The initial mystery that attends any journey is how did the traveler reach his starting point in the first place?

Louise Bogan, 1897-1970, Journey Around My Room

There comes a point I'm afraid where you begin to suspect that if there's any real truth, it's that the entire multidimensional infinity of the Universe is almost certainly being run by a bunch of maniacs.

-Douglas Adams, The Hitchhiker's Guide to the Galaxy

1

Overview

If we could read the thoughts of Aku, leader of a small clan of Sapiens living in the fertile valley of Mesopotamia on the shore of the Black Sea 7500 years ago, it might read something like..

Life is good. It is getting warmer and food is plentiful. For some time now members of my group have heard stories from our trading partners that the Sea Without End, beyond the great hill where the sun vanishes, is reclaiming the land. I indeed have verified this on my recent voyage. I have given much thought to this matter. Perhaps the Sea Without End God is at war with our Sea God. I must encourage all to pray to our God to give her strength and help her protect our valley and our way of life. But the Sea Without End God is very powerful and my clan must reconnect with our old ways and be prepared to move quickly and go to higher ground. If we need to move, I hope our God will give me a vision. It would be nice if the Gods would stop fighting with each other, but I guess that is better than them fighting us. They make life tough sometimes. Some days I am not sure I really believe in these gods; but then how else would I and all I see, be?

Even though the Black Sea God lost and the flood wiped out many other tribes, Aku's tribe did survive. Those genes of Aku's people became incorporated into me. So it is important to consider how Aku came to be, indeed how our universe came to be. How did anything happen?

Becoming addresses Aku's last question. It includes three topics: (1) the origin of our universe; (2) the origin of life here on Earth; (3) the development of consciousness. From the beginning of the Universe to the present there were many possible paths to take; I will take us on that unique path that led to me, an 80-year-old *Homo* s*apien* living in the 21st century. (Following Harai's book, "Sapiens", I shall call our tribe Sapiens.) The overall goal is to arrive at a strong story to explain how and why the events that opened that path came about.

The tested rules and knowledge of present-day science are used whenever possible to suggest a believable story to these questions. Evidence plays a big role. The interpretation of evidence is where people differ. That is why multiple evidence and the observation of predicted events is important for a strong story. A story about past events must lead seamlessly to what we observe at present. If not, that story must be modified or discarded. Often stories of the past lead to events in our observable universe we have not thought to look for and are a basis for intense investigations. I use little or no mathematics but instead rely on analogies and stories.

Science was a great invention; it is pretty recent. There is a reason why the thousand-year period before the 17th century is often called the dark ages. To understand why, Historian David Woolton, in his 2015 book "The Invention of Science" describes the typical well-educated European of 1600 (just about 8 Adult Life Spans ago[1]). I paraphrase his remarks.

In 1600 the well-educated European believes in witchcraft; he believes in werewolves; he believes Odysseus's crew was turned into pigs by Circe; he believes in magic; he believes a rainbow is a sign from

---

[1] (I will use a time measure of 60 years and call it an Adult Life Span, ALS. It represents 60 years, the time difference between 20 and 80 years of age when Sapiens can reasonably be expected to contribute to the knowledge of humanity)

God; he believes comets foretell evil; he has heard of Copernicus but believes Ptolemy; he believes in astrology; he has heard the story of Tyco Brahe's new star observation in 1579 but has not thought much about it; he personally owns 12 books. Little did he dream that 1600 was to be considered the start of the scientific revolution. It started only 8 ALSs ago.

In less than 100 years came the discoveries of Galileo, Newton, and others. Science started on a new road. In the words of Steven Weinberg, *"Scientists of the past were not just like scientists of today ... They had completely different ideas of what there was to know, or how you go about learning it. But the point of scientific work is not to solve the problems that happen to be fashionable in your own day — it is to learn about the world."*

By 1700, the well-educated European had a completely different view; it took less than 2 ALSs. He did not believe those things his grandfather did; he owned hundreds of books. His world view had changed dramatically.

I extended the third section to discuss Myths, Beliefs, Religion, and the Future. Our best guess about the future changes almost weekly; the pace of discoveries and new knowledge is that rapid. The future for a society that soon will be able to design and create humans using emerging micro-biological technologies such as Crispr-Cas9 is amazing, scary, and unpredictable. Some believe that future has already arrived. As Yogi Berra has said, *"The future ain't what it used to be"*. But, I do want to ask, what are the choices?

I do not set out to prove or disprove anything. Science cannot do that. I do seek to arrive at the most robust story based on the observations we have at present.

Over the course of writing this piece, I have tried to critically examine my beliefs, my cognitive map, and to use testable explanations for my stories. Come along this path with me as your guide. Of course we all look at any story with different individual cognitive maps, with connections and associations in our brain map that differ. So beware, I am driving. Here we go. Hold on.

II. Our Universe

The deepest sin against the human mind is to believe things without evidence

-Thomas Huxley

Somewhere, something incredible is waiting to be known

_Carl Sagan

In the beginning the Universe was created. This has made a lot of people very angry and been widely regarded as a bad move.

-Douglas Adams, The Restaurant at the End of the Universe

Can we actually "know" the universe? My God, it's hard enough finding your way around in Chinatown.

-Woody Allen, My Philosophy

2

Observations:

What we observe is not nature itself, but nature exposed to our method of questioning

-Werner Heisenberg

In order to develop any story, we need to get evidence of details pertaining to that story. Evidence is best acquired through direct observations. Collecting evidence and storing it are by no means easy tasks. Images, audio recordings and numbers are all attempts to gather and store objective data. Human memory is a very poor way. The span of time for gathering data by direct observations in our universe is from 380,000 years after the beginning of our universe to the present, a time frame of about 13.8 billion years. Light, across all its frequencies, is the main way we observe. The ability to directly observe at past times is possible due to the finite speed of light. When looking at distant objects, we are looking at past events. There is much evidence that the universe is about 13.8 billion years old and was born with a high temperature and large energy. This has focused attention on both the birth process of the universe and on its evolution to the present. It is not my intention here to give an introduction to Astronomy, but rather to focus on how our Universe came about. For the former I can recommend the excellent nontechnical video course by Alex Filippenko (see references). In the next few pages I describe some events which any strong story must predict. I will also present an overview of the history of investigating events in the early Universe.

Direct observations are possible only from the time when the universe cooled sufficiently to allow atoms to form and light (photons) to freely roam the universe. Before that time the universe was incredibly hot but dark with photons strongly interacting with electrons, protons, and neutrons. The universe was expanding rapidly and cooling. Models, based on known physics, predict that one second after the Big Bang, the temperature of the universe was roughly 10 billion degrees and was filled with a sea of neutrons, protons, electrons, anti-electrons (positrons), photons and neutrinos. As the universe cooled, the neutrons either decayed into protons and electrons or combined with protons to make deuterium nuclei (deuterium is an isotope of hydrogen with one neutron added to the proton).

At about 1 second, protons and neutrons stopped changing from one to the other by exchange of neutrinos. But, neutrons kept decaying spontaneously into stable protons and electrons. Hence, more protons than neutrons were built up. By ~2 seconds the environment cooled sufficiently so that nuclei formation could start. Nucleosynthesis was started. Helium, trace amounts of lithium nuclei, and their isotopes were produced by this time. At the end of a few minutes the temperature cooled to less than one billion degrees and nuclei formation stopped. This fixed the ratio of hydrogen to helium atoms (about 7:1), and to the trace amounts of lithium atoms in our universe. In regions where galaxies have not formed it has not changed over time. We observe these ratios today.

The next event we can observe was ~380,000 years later. We can gather data on this event today. Here is why. By that time the expansion process had cooled the temperature to about 3,000 degrees. The nuclei could then bind electrons and form neutral atoms, hydrogen and helium. These atoms are basically transparent to light, trapping (and emitting) light in certain narrow wavelengths only. (Detection of light in these narrow wavelengths is the way to observe these atoms) So the photons became basically "free". The earliest light we can directly observe is a signature of this event and is called the cosmic background radiation (CBR). The CBR was first observed during my lifetime and now has been extensively studied. For the same reason we can observe only the thin outer layers of our sun, the early universe was simply too hot to let any light out. The light could only escape when the environment cooled to allow protons, neutrons, and electrons to slow down and form hydrogen and helium atoms, the first stable atoms. This stopped the strong scattering and absorption of light particles (photons) by charged electrons and protons and allowed light to be free. This was not a sudden event; it took place over thousands of years. 380,000 years is generally considered to be when this process was completed.

To understand events after this requires you to delve into a good course in Astronomy. The next event, grouping of the matter into stars, was a long time later.

All models of later events incorporate (some say require) the existence of more matter than we have directly observed in the universe. It's called dark matter. So any model of the early universe must lead to the existence of dark matter today. At present nothing is known about the origin of dark matter. It was originally made up in the 1930s to account for the rotation of stars about the center of galaxies. Since then many have studies have looked for dark matter. So far no luck.

The phrase "to understand something" essentially means to have a strong story with lots of direct and indirect observations to back it up. It becomes a strong story when many of its predictions are shown to be correct. Eventually as more and more of its prediction are observed it can become a theory. The bar for having a strong story is pretty high; for becoming a theory it is extremely high. There are many observables that need to be correctly predicted. For example, the observed ratio of hydrogen to helium atoms in the early universe eliminates many stories and has given a timeline for the early universe. A strong story of how our universe has unfolded in time has emerged. Some refer to it as a theory.

3
The Experiments

Nothing has such power to broaden the mind as the ability to investigate systematically and truly all that comes under thy observation in life.
— *Marcus Aurelius*

How can information about events earlier than 380,000 years be obtained? We cannot directly see any light from there. There are two ways. We can try to create conditions of earlier times and then record any events directly. The second way is to use models which predict events in the observable universe that we have not yet observed and look for them.

The observation of helium to hydrogen ratio in the early visible universe is an example of this second method. For observations one looks at the characteristic emission lines of the atoms (and their isotopes) in isolated regions of our universe to deduce their numbers. The general number of helium atoms has increased over time due to nuclear fusion in stars which converts hydrogen to helium. So, looking at or near stars will not work. Instead we examine the regions far removed from any star, regions where no star has yet formed. Fortunately, there are many such regions. These observations started in the late 1930s and the ratio is now well known. Physics tells us this is the hydrogen/helium nuclei ratio that was fixed when temperatures dropped below about 1 Billion degrees C (modeled to be 2 or 3 minutes after time zero), and the fusion process of making He-4 nuclei started. By the time of 15 to 20 minutes after it all started, the ratio of hydrogen to deuterium, He-3, He-4 and Li-7 all became fixed. They remain that same even today. The number of helium and hydrogen atoms and isotopes are deduced by the observed strength of their characteristic emission light. Details of the fusion paths are well known. These ratios have been used to whittle down the number of possible stories of earlier times.

Events in high temperature regions have been directly studied for almost a century. The invention, by E.O. Lawrence, at the University of California Berkeley, of the cyclotron was the first technology to allow information from this temperature region to be accumulated. Over the years from 1931 to 1948, researchers at Berkeley improved the cyclotron design and reached an energy of about 100 Mev (Mev is short for million electron volts and 1 MEV is a temperature of about 1, 000,000 degrees). By the mid-20th century we had some pretty good models as well as a bunch of data in this temperature region.

This has led today to the Large Hadron Collider (LHC), the European high-energy machine. In June 2015, it started to operate at twice its previous energy and can now create conditions (650 trillion degrees) when the universe was very hot and dense. This energy region is way above that needed to initiate fusion and I remember cries of caution; warnings that we could create a run-away black hole that could eat the whole earth. This myth soon received no traction as it was pointed out that cosmic rays hitting the earth every day have energies up to 1000 Tev and do not trigger such an effect. There is no shortage of myths in today's world.

Guided by models, two chief experiments (Atlas and CMS) probe conditions of earlier times. The object is to study further details of the previously observed Higgs field (often called Higgs particle) and hopefully to observe new particles.

To create the high temperature (energy) environment the LHC smashes high speed protons together. It accelerates two beams of protons traveling in a circle in opposite direction to very high energy, and arranges for them to hit each other. Detectors are set up to just observes what happens. The energy of motion of the protons is transformed into heat energy, defined by a high temperature, in a very small region.   An analogy describes the process as trying to investigate the inside of a fine Swiss watch by hitting the watch with a hammer and observing the pieces produced. A more powerful hammer can show even finer pieces to examine. There are three main differences between the environments produced in the proton-proton collision case and the watch-smashing case. In the first there is a strong repulsive force between the protons that gets stronger as they approach, the products last only in the high temperature environment (so one needs to look very quickly), and lastly the experiment needs to be repeated many times and allow the statistical footprint of observables to build up. It may sound simple to arrange but, as often, the devil is in the details. The LHC was built below ground to isolate the experiments from things continually hitting the earth's surface, events that cause "noise" in the observations. It was built in a circle 27 kilometers in circumference; there are 1600 superconducting magnets each weighing about 27 tons keeping the beam inside an incredible low vacuum ring. Added to that construction task is the design and fabrication of detectors to look at collision details. To create a very small environment of close to a billion degrees requires parts of the apparatus to be at temperatures close to zero degrees (absolute degrees, in Kelvin).

The LHC produces about 30 million billion (30 x 10^{15}) events that must be analyzed each year. That results in over 5 times the largest data base we have ever had on earth. Data analysis alone required advances in super-computers.

The second method to gain information from the early universe is almost as difficult in practice as the bruit force method. The Cosmic Background Radiation was first observed in 1941 at Bell Labs here in the US by Penzias and Wilson. They had a project to investigate the limits on transmitting and detecting radio waves. Their experiments were shut down for the war that came shortly. After the war the work slowly resumed with development of more sensitive antenna and detection designs. In 1964 they had done their best and still could not get rid of "noise" which seems to come from every place in the sky. This was quickly realized as the cosmic background radiation predicted by a number of ideas about the start of our universe. Since then more detailed observations have revealed details of the small changes in the CBR over the entire sky and have already eliminated and refined more stories.

The experimental goal is to observe small differences in the Blackbody Radiation over the entire sky. The 3,000-degree temperature of the CRB is red-shifted to the microwave region; the radiation peak determines its temperature of 2.76 degrees absolute (Kelvin). This method requires a large radio telescope to collect and then focus the radiation onto a detector. That is the easy part.

Most all of the development effort for looking at the CBR concentrates on the development of highly sensitive detectors to resolve small details in the 2.76 degree "black body' radiation. What is required are observations that detect a few micro-degree change difference; in practice detecting a difference of few part in a million in an already very weak signal. It is a difficult job to create such detectors but already some have been developed that have observed small regions of temperature change of the CBR across the sky. This translates to a difference in density of atoms across the early sky and helps refine timelines and models of galaxy formation (galaxies started to form at about 200 to 500 million years). In addition, last year refined data of the CBR showed hints of rotation of the polarization of the light. All this is just what one of the popular models of creation, the Inflationary Hot Big Bang Model, predicts. This model predicts the inflation period existed in the period 10^{-36} to 10^{-32} seconds after creation, times well before the LHC can even dream to achieve.

Detectors have recently increase in sensitivity, and soon finer details should be available. If polarization rotation is verified, it will make the inflationary story much stronger.

Polarization of light may need some comments. It has been extensively studied for years and details of polarization are well known. Light polarization can be easily observed here on earth. On a sunny day, take your polarized glasses and look through them at any patch of sky 90 degrees from the sun. This angle makes the sunlight, scattered by the atmosphere, polarized. Your glasses allow only one direction of polarized light to pass. As you rotate your glasses the intensity of the polarized light can readily be observed.

So far, the research results from both of the investigations described here, have led to a creation story that is getting increasingly more robust. The story is almost unbelievable; no science fiction writer could dream it up. The model implies that our universe came from nothing. It was a random event. How can this be when in our present environment things seem to be governed by cause and effect? The universe obviously has a lot of energy. How did that happen; what is the story?

4
The Creation Story

Man has been here 32,000 years. That it took a hundred million years to prepare the world for him is proof that that is what it was done for. I suppose it is. I dunno. If the Eiffel tower were now representing the world's age, the skin of paint on the pinnacle-knob at its summit would represent man's share of that age; and anybody would perceive that the skin was what the tower was built for, I reckon they would. I dunno
--Mark Twain

Other than his numbers being off a bit, Mark Twain hits the nail on the head in his own satirical way. The Universe was not made for us *Homo sapiens*. We arrived very, very late. But was it made for life? How did the Universe come to be? What are the stories and their predictions that scientists are trying to prove or disprove? What is the best story? Based on direct and indirect observations, a number of stories have been eliminated and now there is a single serious story, a fairly robust model, for the beginning of our universe.

To get some understanding of this model, I use an analogy: the whole integer number system; 1, 2, 3,…. and its counterpart, -1, -2, -3,…, written here in the Arabic system used almost exclusively throughout the modern world. The whole number zero, 0, was a great addition to the set of positive and negative whole numbers. Zero plays different roles; for example, when zero is used on the right side of a whole number as in 2 to get 20, it serves to denote a power of 10. But here I will concentrate on its use to mark a distinction between the positive and negative whole numbers.

Most will agree that a +1 plus a -1 would add to 0 and can be represented as (+1) plus (-1) = (0). But, can we go the other way and get (0) = (+1) and (-1)? Of course one can do this with numbers, but can that have anything to do with reality? Can we get something out of nothing? Can we do this multiple times?

Least you think this question is just a play on words and cannot have any connection with the real world, consider a dream bank that gave you a loan of $1,000,000 with no interest for one year with no collateral required. You had to sign a promissory note and leave it with the bank. You have just taken nothing and broken it into (+1) and (-1) but for only one year (actually into +$1,000,000 and -$1,000,000).

Now suppose you get into a space ship traveling at the speed of light; further communication between you and the bank is now no longer possible. In your universe of the spaceship you have just created something out of nothing. The bank is left behind with only a lot of potential. Sounds almost like a Ponzi scheme. What about the rule of not going at or faster than the speed of light? Early in the 20th century the consensus that nothing in our universe goes faster than light was proposed, tested and found to be correct in all cases considered. Special relativity is based on this assumption and there are many experiments verifying this.

However, this speed of light limitation refers to things in a space and imposes no limitation on the expansion speed of space itself. It can go any speed. So, there is no contradiction in having space expand at faster than light speed. Keep this in mind as we develop this idea further.

Analogies can be fun and worthwhile to explore; however, we must remember that at some point they may not relate to the original situation. Eventually we need to make connections to that original problem. We were at that point near the end of the 20th century. Folks knew that quantum theory, namely the famous Uncertainty Principle could be used to supply any energy needed. This last concept is at the heart of quantum theory and is well verified. Using energy in place of money; the larger the energy borrowed, the less time you have to pay back. It is a form of the famous "Uncertainty Principle".

So, how does it work and what can it predict? The details of this getting something for nothing concept starting our universe was put together in the late 1980s by Alan Guth during his 8-year post-doc days when he was also looking for a tenure track university position. His formulation shows how you can in effect get on a spaceship and not have to pay back the "loan". It is called inflation theory. It is a bit mathematical, but by no means just a wild hunch. An early inflation period is consistent with all the observables in our universe, the big first step of any model. The success of this model has led to the concept of multiple universes, using the argument that if it happened once it surely also happened a number of times. This last thought of the existence of multi-verses borders on science fiction as it makes no observable predictions (yet!). None-the-less the lack of testable predictions does not stop theorists from dreaming about multi-verses. Here is one such speculation.

In a recent (2016) paper the CRB over the entire sky was calculated based on the best present day model. When this calculated CRB was subtracted from the actual data, points of light remained. The authors suggest this could be the result of collisions with another universe born near or at the same time. Presently this is, of course, science fiction, but beware science fiction has a way of sometimes becoming true.

 A universe from nothing is certainly a strange reality, but I believe a correct one. It is formally called the Inflationary Hot Big Bang Model. There are still many details of the nature of dark energy and dark matter to uncover, but many, including me, believe they can be understood within the framework of this model.

 Models themselves often lead us to new concepts, and a completely different way of thinking. A good example is the concept of Gravity. It has changed from the days when Newton showed that his Law of Gravity could describe both an apple falling from a tree and the falling of the moon around the earth. Newton himself was puzzled by this "spooky action at a distance" rule for the gravity force. His critical question was "How did the moon know what the mass of the earth was?" Little remembered or appreciated for over two centuries is the description appearing very soon after Newton's by the French mathematician, LaPlace. He introduced a gravitational field existing throughout all space as a result of a mass M and showed that the gravity force on any mass, m, resulted from spatial changes in that field. His work introduced the idea of fields and resulted in the exact same force as Newton's. So the response to Newton's critical question is that the question no longer makes any sense when you think in terms of fields. Only recently, in my lifetime, has it been realized that fields are the fundamental way to look at all forces and that LaPlace's mathematical description was indeed a great conceptual development. I might add again that science fictions writers were quicker than professional scientists to grasp the reality of fields.

The present day answer to the long debated question of whether light is a wave or a particle is now settled. It is a wave. The photon is a disturbance in the electric field that travels at the speed of light. An analogy would be a "jiggle" traveling along a stretched rope. Similarly, the Higgs particle is a disturbance in the Higgs field and the graviton a disturbance in the gravitational field. Acceptance of fields has led to a new approach to understanding the make-up of our Universe and the use of mathematics to guide the search for understanding.

Yang and Mills have pointed out the value of using specific symmetry groups as a tool in physics; this resulted in what is now called the standard model. This model was verified (it has passed from a strong story to a model) with the discovery of the Higgs field (often referred to as the Higgs particle) in 2014.

It continues to be amazing that abstract math can describe real world phenomena. As described by Edward Frenkel in his wonderful little book "Love and Math; the Heart of Hidden Reality", Nobel Prize recipient Yang pointed out in his acceptance lecture; *"It was not just joy, something deeper. After all, what could be more mysterious, what could be more awe-inspiring, than to find that the structure of the physical world is intimately tied to deep mathematical concepts, concepts which were developed out of considerations rooted only in logic and the beauty of form?"*

Einstein, decades earlier, expressed similar thoughts; "How can it be that mathematics, being after all a product of human thought independent of experience, is so admirably appropriate to the objects of reality."

Today work beyond the standard model uses similar mathematics to develop theories such as string theory and super-symmetry theory to guide experiments.

Although important details such as the origin of dark matter and dark energy still remain largely unknown, the general story of the creation and evolution of our Universe over the past 13.6 billion years is pretty clear, and so far, well documented in studies. In 2016 the LHC started to take data at very higher energies. Preliminary data shows the presence of a particle with six times the energy of the Higgs. If verified by further data, this narrows the list of models and most likely will point to new physics.

All this creation from nothing story, called now a theory by many, may seem to be just a wistful dream to many. Descriptors other than wistful, such as impossible, ridiculous, brilliant, have also been used. To become a theory, a dream must become a strong idea which has many predictions that can be tested and verified by observations.

Humans have long thought about our universe. I imagine our early human cousins, the Neanderthals, sitting in the evening by their protective camp fire, made up stories about those points of light in the sky they could see. They did not have street lights, cell phones, or TV to distract them. Their observations were that the earth was still and there was a sphere that revolved about it. The sun and moon moved about on that sphere and, if they looked carefully, a few objects on that sphere did also.

We fast forward to the times of Copernicus, Bruno, and Galileo. Their dreams and observations were not treated well. Certainly Bruno's was not. After being charged with heresy in 1593, Giordano Bruno was burned at the stake by the Catholic Church under Pope Clement VIII in 1600. The charge was continuing to teach the universe is infinite and that we are but a small part of the universe. Bruno believed there were other worlds, some with humans like us. It was a poor time to be alive and to be an outspoken free thinker.

I have visited the old Roman market square (now called Campo de' Fiori) in the center of Rome Italy, where the killing event took place. I learned that Bruno's mouth was taped shut so he could not speak as he burned to death. Even today some respond to the question, *what is truth?* with the reply *Truth, truth is what Bruno knew, but could not speak.* Bruno is one of my heroes.

Three hundred years later free thinkers were more tolerated. Today, Sapiens even use the word treason instead of heresy when the state kills or incarcerates someone who speaks against the state, thinking that word changes justify the action.

In 1915 Einstein's General Theory of Relativity, called a new theory of gravity, was published. It was very mathematical and its predictions relied on the belief that mathematical models can predict observable results in our world. The early version of this model relied on a few basic assumptions but showed that the theory did not predict a static universe filled with matter. So, he just added another term to his equations and adjusted its strength until he got the prediction of the universe most agreed with, at that time an infinite static universe. I like to believe his published paper would not have survived the peer review system science has in place today. You simply do not make a result an integral part of a theory which then goes on to predict that result as proof of your theory.

In 1922 the Russian mathematician Alexander Friedman published solutions of Einstein's original equations with the simplifying assumption of a universe uniformly filled with matter. When he varied the starting conditions, he discovered most of the solutions give a universe either expanding or contracting. Only under very special initial conditions did it lead to a static universe. No one seemed to pay this result much attention; after all, could anyone believe in a theory that predicted an almost zero probability that we are here? When Friedman died at age 37 three years later in 1925, his results were quickly forgotten.

The year Friedman died, Hubble's observations showed we live in a universe of galaxies. Two years later in 1927, Lemaitre independently discovered Friedman's results. Lemaitre was a Belgian priest who was a PhD student at MIT at that time.

In 1929, Hubble's observations showed our universe is an expanding universe and Lemaitre work suddenly became of interest. Hubble had also measured the rate of expansion of the universe. When Friedman's equations (now called Friedman-Lamaitre equations) were run backwards in time, the prediction was that everything, the density, temperature, etc. goes to infinity. No one likes infinities. For 40-50 years, little progress was made to develop models with no infinities. However, data strongly supports a beginning. Many in the science community thought that belief just wrong; there was no beginning. Just to emphasize how absurd a model with a beginning was, it was soon called the Big Bang Model by people who disliked it.

They came up with a continuous creation model, jiggered up to give a number of observable results. When I first taught Astronomy in the 1970s, all the textbooks covered both models, but really only one deserved to be called a theory.

About 1980, Alan Guth came up with the idea of inflation and showed it is allowed by Friedman's equations. To start he knew a quantum energy fluctuation in time could give any energy that was needed. So his big dream starts with a small mass (energy) which doubles in step with a doubling of space. This process is exponential. Max Tegmark, in his book "Our Mathematical Universe" gives us numbers predicted by this theory of inflation and describes an analogy in terms of a baby's development.

If a baby, after conception, doubled its number of cells every day, it would increase in number as 1,2, 4, 8,16, and so forth, but the density, of the perhaps not quite yet baby, would remain the same. Of course energy is needed to do that. This exponential growth has been observed in humans and is believed to occur at least for the first three days of development. If it continued for the 270 days of pregnancy, the baby would grow to much more that the mass of the universe and your mother certainly would not thank you.

To reach the universe mass, about 260 time units for doubling does the job. Although detailed models of inflation differ, to get our universe most predict doubling every 10^{-38} sec. and an expansion lasting 10^{-35} sec. The energy would come from a quantum fluctuation. With the result of experiments described previously, the overall inflation model is fast becoming a theory and we can marvel again at the connection between pure mathematics and reality.

Before you start thinking that science is wonderful and that we are close to a real understanding of how everything came to be, I need to describe a few things, among many, we do not understand. Much of our understanding about very early time relies on data in our observable universe predicted by mathematical models of earlier times. When our present theory of gravity makes predictions at earlier times, it does not agree with "the standard model" observations. Plus, it predicts infinities!

Right now "the standard model" is a quantum field theory with inflation that includes three of the four forces we observe; gravity not included. Starting with early work by Einstein there have been many attempts to develop a quantum gravity model. No success yet.

We need a field theory that does not predict infinities at small scales.

Then, there is the "fine tuning problem." Some constants of existing theory need to have very precise values to predict the universe we have now. We have no idea how these constants came to be but very small changes in any of about 18 of them and I would not be here writing, nor you reading.

So, how about inflation itself? We do not have any information on how it is turned off, but there are a number of ideas on what the universe was like before expansion. Prominent among these is string theory, which really should be called "string ideas". String theories are a way to avoid the infinities generated by General Relativity as distances go to zero. We get rid of the infinities simply by saying we never go to zero. Sounds like a cop out, but it is really a serious idea. This idea is that everything evolved from strings and strings take a finite space. When the math of this idea is carefully done more than 4 (three space and one time) dimensions are required. Right now the favored math model is one requiring 10 dimensions. It seems to lead to what is observed thus far but also predicts supersymmetry particles which should be observable today at high enough energies. (According to all models there is no feasible way to reach the extreme energy region of strings or even quarks.)   Presently the LHC experiments are looking for signatures of these predicted supersymmetry particles. There is a very readable discussion of string theory at superstringtheory.com (see reference list).

Remember String theory is a way to ensure gravity interactions via strings take place over a finite distance, abet a very small one. It needs to be compatible with the standard model and also, both need to lead to all the observations thus far. A tall order, but some progress has been. String theory is associated with supersymmetry and a multi-dimensional universe. Max Tegmark, in his recent book "Our Mathematical Universe" has more to say on these ideas. But remember they are so far just ideas even though they go by the name theory.

Not yet settled questions that have been around for a long time include the following two: (1) why is it that the gravitation and electric forces each vary as one divided by exactly the square of the distance? Both have been measured to such an accuracy that we can say exactly; (2) why do we find positive and negative electrical charges but not positive and negative magnet ones? String theory may be able to enlighten us on these questions.

There are other on-going experiments today that may shed light on how our universe came to be. One model of dark matter suggests a weak interaction between normal matter and dark matter. The concept of dark matter first came about when we discovered there is not enough observable matter in galaxies to explain their rotational motion. Dark matter interacts with normal matter via the gravitational field but only interacts very weakly, or not at all, with the other fields. Dark matter particles, called dark since we cannot see them, are called weakly interacting massive particles and are commonly referred to as WIMPs. There have been a number of experiments to detect WIMPs, none successful. The latest, named XENON1T and housed at the Gran Sasso National Park in Italy, sits in a cave 1500 meters below ground so as to shield out noise. It uses a very large amount of liquid xenon as the normal matter. In late 2015 experiments were again started at this facility; this time with enhanced sensitivity light detectors. No WIMPS yet.

The recent detection of gravity waves attributed to the collision of two fairly massive black holes points to another method to study gravity in a hot dense environment. This event was detected by the clear identical signals observed at the same time by two similar facilities in the US, one in Hanford, Wash., and one in Livingston, La. The small time delay of 7 miliseconds in the signals allowed a source location in the sky to be inferred, but not identified. These are called Laser Interferometer Gravity-Wave Observatories (LIDO). There are worldwide plans to create many LIDOs, with the hope of discovering new physics and to pinpoint the location of the source and should verify the speed to gravitons and give more information on the nature of the gravitational field. More LIDOs will be operational this year. Further data should lead to better models of the early universe. To get some understanding of why this is an important observation we need to look a bit closer to various field theories. August 2016 thoughts on these data can be seen in Emily Conover's article in Science News.

Sean Carroll in his 2013 blog (see References) has a very readable piece on constructing quantum field theories. The process in principle is simple: start with a field theory, like the electromagnetic field we have seen previously, apply the standard rules of quantum mechanics and, voila, a collection of particles; a quantum field theory results. This allows one to make calculations about how particles interact and to make predictions. Richard Feynman showed how to organize the interactions in a pictorial way, now known as Feynman diagrams, so as to facilitate calculations. Unfortunately, some of the contributions give infinite contributions. If one ignores these contributions, exquisite agreement with measurements is obtained. Initially this was called renormalization. Further work by Wilson and others showed that these infinite terms appeared only at higher energies and can be safely ignored at lower energies; this is why the LHC pushes into a higher energy realm to see what new physics waits to be known. The LIDO discovery is already important; it shows that gravity can be described by a field theory that somehow can be renormalized at high energies. So far no gravity theory can do this. Fields like the Higgs Field and the Electric (actually the Electro-Weak) Field are fields that exist imbedded in space-time but a gravitation field is space-time itself. So it is a little different. Look for ideas of Quantum Gravity, Superstring Theory, and Supersymmetry to flood the literature soon. There is no lack of ideas. One getting some serious study relates quantum entanglement and possibly virtual particles with wormholes predicted by general relativity.

The quest to understand continues and, yes, science is never straightforward, often messy, often expensive, but also beautiful and playful. Besides, it is quite enjoyable to spend time in places like the mountains of Italy, but perhaps not so much in Hanford WA., Livingston LA., or 1500 meters below ground.

Let's take a second look at one of the descriptors I have used here ... "a Universe from Nothing". Many would argue that it is impossible to get anything from nothing. There had to be something to start with. In these seemingly opposite views, the word nothing must have different meanings. In the first case nothing means no space and no time. In the second case nothing means creation within a space-time space and science agrees there must be something to cause this. Fields are there in the vacuum. Fields that have energy and effects that can be measured. Virtual particles are created all the time in accordance with the quantum uncertainty rules discussed earlier. They pop in and out of existence quickly. This story has observations to support it. Creating space (actually a space-time space) is different from creating something imbedded in an existing space time.

Now on to the second topic on my list and examine the information we have to develop an informed story. Here, unlike in the present topic, we can get some information at times before and after the event, and hopefully use that to guide the development of our story.

III. Life

Modern man may yet turn out to be the missing link between ape and human beings.

-*Stuart Caesar, Philadelphia, 2006*

If aliens visited Earth, they would take some note of humans, but probably spend most of their time trying to understand the dominant form of life on our planet - microorganisms like bacteria and viruses.

-*Nathan Wolfe*

5
Life on Earth

Evolution is cleverer than you are

-Francis Crick

Evidence shows that life began on earth about 4 billion years ago; it appeared on our planet shortly after the earth condensed out of the dust and gas mixture that formed our second generation solar system. It formed almost as soon as water appeared. Water was essential for life to start. Over a short span of the last 100 years many of the details of life have become understood.

During the period 1951 to 1953, the structure of DNA became known. This discovery changed our understanding of life. All life we have found and studied so far is based on that DNA left handed double spiral amino acid chain structure. Many identical pieces of the same long DNA molecule are found in all forms of life today. To get a picture of this structure imagine a rope ladder with two long strands with wooden sticks connecting them in many places. The wooden sticks are made of two smaller sticks, matched pairs of amino acids. There are held together by pretty weak chemical bonds that stick together in the middle. The other ends are attached to one of the ropes. There are only two types of sticks requiring only four of the 20 amino acids found elsewhere in life. Sugars, proteins, and phosphorus join together to make up the ropes. A number of scientists contributed to the definitive paper in Nature, April 25, 1953, announcing the structure of life. It was authored by Watson and Crick.

Shortly after the DNA structure became known, Stanley Miller then a chemistry graduate student at the University of Chicago lobbied his thesis advisor to let him think and study ways to get the four essential amino acids needed for DNA from the components existing on early earth. He was given 6 months to work on this project; it was considered a long-shot to produce a worthy publishable thesis and the role of a thesis advisor was to not guide students into a dead end project. To replicate early earth conditions Miller used a mixture of methane, water vapor, ammonia, and hydrogen gasses. He put these together in a beaker, varied the temperature and cooking time, and found nothing of interest. Miller then decided to simulate lightning, as it was thought also to be present in the early earth. A high frequency voltage between two wire electrodes would cause sparks in his beaker. After a few days of running his new set-up, Miller found brown gunk coating the beaker and electrodes. Analysis showed the gunk contained amino acids.

Newspapers shortly after publication of Miller's work, boldly proclaimed **Life Made in a Test Tube.** Of course that was not the case at all. It was a good lesson to not believe all you read or hear, but instead go to prime sources, like Miller's peer reviewed publication (unfortunately, every 4 years US national elections remind us that many refuse to believe that lesson). All the Miller experiment claimed and showed was that it is relatively easy to make amino acids. It is also easy to make (synthesize is the chemist word) proteins, and the sugar molecules which, when joined to an amino acid in the right way, and these joined together, make one side of DNA.

But, how these all get cooked by chemistry, presumably in a warm pool of water as Darwin supposed, to form life remained unknown. I remember the words *primordial soup, warm shallow ponds* and *somehow* used often in the stories about that process. But making life requires much more than making amino acids and these magic words.

Life itself is difficult to define. Andrew Knoll, a current Harvard Biologist defines it this way:

"I think you can say that life is a system in which proteins and nucleic acids interact in ways that allow the structure to grow and reproduce. It's that growth and reproduction, the ability to make more of yourself, that's important. Now, you might argue that that's a local definition of life, that if we find life on Europa at some time in the future, it might have a different set of interacting chemicals."

As we will see in Chapter 7 when we look at viruses, there are other definitions of life.

6

Bacteria and Archaea

We inherit every one of our genes, but we leave the womb without a single microbe. As we pass through our mother's birth canal, we begin to attract entire colonies of bacteria. By the time a child can crawl, he has been blanketed by an enormous, unseen cloud of microorganisms - a hundred trillion or more.

-Michael Specter

Bacteria, single cell life with no nucleus or other internal compartments but satisfying Knoll's definition, were thought to be the oldest life on Earth until the year 1977. That year, deep sea explorations observed abundant single cell life near sulfur vents a mile deep in the oceans. This new life relied on sulfur rather that oxygen for an energy source. It still had the double helix structure. They were called archaea, meaning ancient life.

Archaea exist in a variety of forms. They are more evolved than bacteria and are considered older. They are thought by many to be the earliest life form and, along with bacteria, make up prokaryotic life (a cell with no separate internal compartments such as a nucleus to house DNA). During the past 40 years, archaea have been found in numerous places including surface soils, marchlands, and the human gut. **Marine plankton, at the bottom of the ocean food chain, show about 20 % archaea DNA components.** Some archaea even use oxygen as an energy source.

As 3.8 billion years ago (BYA) these two distinct forms of early cells that could reproduce by cell division, dominated on earth: bacteria and archaea, each with many different sub-structures. These competed for survival on early earth. Both archaea and bacteria contain no defined nucleus and no well-defined membrane-bound molecules inside their cells. Both cells have DNA, RNA and proteins that supply everything necessary for survival in a supportive environment. They are still here today. We study them. With the discovery of archaea, the question now asked was could sulfur, water, a mix of the early oceans environments (which had changed a bit from Miller's day), pressure, and perhaps heat, form amino acids. Could we substitute high pressure for a lightning discharge?

That experiment was tried but failed; this led to the question of what else was present in the early earth environment that may substitute for Miller's electrical discharge. This quest was led by a team of geologists and they knew that ground up rock was very plentiful on early earth (as was sulfur, both due to volcanic eruptions). With ground up rocks mixed into the "soup" and the experiment run, analysis showed not only amino acids, but also a number of other proteins had formed.

Fine ground rock is what forms mud and then clay. It is found at the bottom of ponds. It forms in two dimensional sheets. Mud can be pictured as layers a few molecules thick, separated by water. There are many surfaces. Surfaces are well known to act as powerful catalysts, accelerating chemical reaction rates many, many times.

How about availability of other molecules needed to build up strands of RNA and DNA? As reported in Science April 2016, when ice with components of water (H_2O), methane (CH_4), and ammonia (NH_3), are exposed to UV light and allowed to come to room temperature, a variety of sugars, including ribose ($C_5H_{10}O_5$), form. Ribose is a key component of RNA and DNA. Ices containing these three molecules, water, methane, and ammonia, are thought to coat grains of dust about early stars; the material which condensed to form planets. They have also been found in comets. A number of other molecules important in life, including many amino acids have been observed in space. So on early earth there was a plentiful supply of molecules which are found in all life. All that was then required was time to try out different chemistries. It then seems plausible to me that the basic building blocks of life came into being in that warm pool that Darwin proposed and that life appeared all over the planet, basically at the same time. Water protected these early molecules from the daily UV radiation of the sun which tends to break them apart, and mud layers provided the surfaces to speed up reactions.

But wait, all the amino acids as well as 95% of proteins found in all life are left handed. In the Chemistry lab we always produce equal numbers of right and left handed amino acids. (place your thumb in the direction of the helical spiral, your fingers of either your left or right hand will then curve around in the direction of the spiral). How can we get an abundance of one over the other? Recall that life had to start under water to protect its molecules from being broken apart by UV light. This light can be polarized by scattering off molecules, whether in the atmosphere or in water. Reflections from surfaces change the handedness of the light and can lead to a difference in the strength of right and left hand polarized light. This light is absorbed more readily by molecules having the same handedness and could give the advantage to left handed biological life. Now we do not know that this happened, but it is a reasonable guess. If we do find life elsewhere in our universe, one of the most interesting questions is to ask about is the handedness of their DNA and other biological molecules.

It is believed by many this pre-archaea environment acted as a "chemical lab on steroids", building up and testing many strands of RNA, joining them, and then forming the first cell.

7

Viruses

There is nothing so patient, in this world or any other, as a virus searching for a host.
— *Mira Grant*, *Countdown*

It took a long time for archaea and bacteria to evolve from amino acids and sugars. A lot of biochemistry experiments went on. We will examine what is known from present day microbiology studies as a guide to help understand some of these details. That is just what has been done in the quest to understand how the universe came to be. Can present day knowledge point to a path or paths leading to bacteria and archaea There are many steps of interest to consider. In this chapter we visit the world of viruses and their possible role in early life development. Some scientists say viruses aren't alive because they rely on their host cells to reproduce. But a number of scientists disagree. A not-life argument is that they don't carry with them everything they need to survive. But neither do bacteria or humans. What's important is that viruses transmit genetic information from one generation to another. With this view life is information that replicates. On that basis viruses are life.

Virology is the study of viruses. A virus is an organism about 100 times smaller than a cell. All viruses have genes made of short strands of either DNA (double stranded) or RNA (single stranded) helical molecules. The length of these structures vary from about 10 to 300 nanometers.

For many years it was thought that all viruses seemed only to attach themselves to the surface of a cell, burrow in, splice itself into a cell DNA, and use the cell machinery to make copies of itself. So they certainly can reproduce and pass on genetic information. However these viruses needed to use another DNA and hence they were classified as not self-reproducing and non-life. In the next chapter we will see that view is changing.

Viruses can inflect all type of cells, almost any cell will do whether in animals or plants. These viruses do not carry their own energy source. Once outside a cell they can be disassembled either by changes in their environment, predominately temperature and pressure changes, or by attacks from other viruses. Other than being pulled apart, they just become dormant, not dead because they were presumably not alive to begin with. Some remain inactive over a wide range of environmental conditions for well over a year. There are many different types of viruses and they are found everywhere. It is estimated as many as 900 million viruses may occur in one milliliter of seawater and there are 10^{31} viruses on earth. Viruses, particularly RNA viruses can mutate rapidly in a cell; they carry no proofreading mechanism for those genes when the splitting and recombining necessary for cell division occurs.

Did viruses or bacteria cells come first? You may think this is like the chicken and egg question that we all heard as kids and dismiss the question; but look closer at the difference between early cells and viruses. Of course both may have evolved and we really need to imagine them as they were 3 or 4 billion years ago. However, to start assume the oldest simple cells are like the general bacteria cells found today. Inside a fluid structure, protected by a cell wall, there is a dynamic region of circular DNA plus proteins and RNA molecules. Throughout the cell there are many smaller regions containing DNAs of only a few genes. These are called plasmids. The DNA in both places can reproduce and the regions can move about. Nothing is particularly constrained to stay in their region but they do most of the time. The first question will be where the energy necessary to run the cell comes from. A quantum fluctuation followed by expansion as in our previous story is not applicable.

If this energy source is not contained in the cell, it must be taken from outside, and the ultimate source of energy must be the sun. The sun's energy could be directly put in to the cell, or it could go through an intermediary storage phase, in chemical compounds made available in the environment. We have found present day bacteria that do both. I believe breaking up chemical compounds, which requires the presence of enzymes, was the original mechanism to get energy from the environment. It was simpler and most researchers believe photosynthesis came later. In any case, there is no problem getting energy.

What do recent studies have to say about early life? What are the ideas and what do the experiments say about these ideas?

8

Recent Thoughts and Studies

I was flabbergasted at the result, I couldn't believe it -Dr Martin

Now we can dream, generate some ideas, and see what the evidence is. Here is one: In the early Chem. Lab on Steroids, many compounds were tried; most broke apart quickly. Label these as unstable. Any structures we can find today are most likely stable ones. Could small structures of atoms join together to form larger ones and the process continue until some stable or semi-stable molecules form? Here we must enter the world of Chemistry and ask what the electrical force can build with atoms and molecules. What can be made in the Chem. Lab on steroids?

Here is an idea. In the "eat or be eaten" environment of their world the molecules found themselves in need of a protected space. The small DNA in plasmids developed the ability to code for molecules that produce antibodies so as to kill other viral invaders. A small wall was built around the plasmids and some of the things they needed so as to make it more difficult for the viruses to get in. The antibodies could then interrogate the potential bad guys outside the wall. If they recognized them as bad, they could just pull them apart; otherwise they just let them drill into the cell, maybe they could do something useful. This provided a safe place to try out new DNA by incorporating roving strands of RNA or even DNA. These plasmids could explore even more chemistry. Many different trials at getting larger and making an even more protective walls went on. Most of them either reached an unstable state and fell apart or evolved to a dead end state, just being inactive. Eventually the road that led to the first simple cells, cells we observe today.

This guess makes a number of predictions; it implies a number of paths taken, but many not successfully. So what happened to those small amino acid and sugar pieces that stuck together and went down dead end paths? Were they just pulled apart and sent backward on their evolutional path, perhaps to the start, or did some reach the end of their evolutional path and find a place to survive, perhaps to just sit there, hiding in plain sight? Is there any evidence this idea is actually a story or a strong story?

The discovery of the first mega-virus came about by chance. In 2002, T Robotson collected water samples from a hospital cooling tower in Bradford England. There was interest in infected bacteria running around hospitals at that time and his goal was to collect bacteria and see what they were like. He found one that had a strange DNA sequence unlike any known in the bacteria world. It was as large as bacteria, had 1018 genes, but had a round structure that was usually found in a virus. This was reported to a collaborator in France who then found a similar structure in a French hospital water stack. Since then these strange structures have been found in many places; in sea water, in animal guts, and even in 30,000-year-old frozen tundra in northern Russia. They can replicate themselves and are sometimes called virus factories. Of course, these giant viruses could be attacked by smaller viruses so they most likely developed some type of protective environment. They seem to fit the requirements of life set forth by Knoll. Are they life? Did they evolve to a stable dead end and are just there doing nothing? How did they survive the viruses attack? The first structure was called Mimivirus (for mimicking a bacteria). It is interesting that these large life forms, or almost life structures, were only discovered recently. The reason may be that the standard way to look for viruses is to first filter out from your sample all the stuff that is "obviously" too large. So these large viruses were excluded and discarded before analysis even began.

In 2010 a French team led by Jean-Michel Chaverie discovered (in water) an even larger virus with a DNA of 1259 million base pairs and 1120 genes. They called it Megavirus When its genes were compared to those of Minivirus (1181 million base pairs), there were direct matches for 258 genes (including those that coded for the viral factory), however there ware family matches for most all genes. Both viruses code for sugar, lipids, and amino acids. Another virus of note is the smaller cafeteria virus, 730 thousand base pairs with about 544 genes. It also codes for the viral factory but shares only about a third of genes with Minivirus.

Scientists at the Craig Venter Institute in La Jolla, Ca., a private biotech firm, reported building bacteria with minimal genes of 474 with the function of 149 genes unknown but essential for life. This was announced March 25 2016 in *Science.* These are built from the bottom up and are called synthetic life; they seem to satisfy all the requirements of life.

This story presents a straight time-line story with many dead end branches, leading from the first amino acids to viruses to bacteria, to cells we see today. The last three paragraphs may describe intermediate steps, but maybe they are not dead ends. Maybe they evolved into different types of bacteria. Over the past 15 years more than 1,000 new types of bacteria and archaea have been identified, hiding in their own environments. So far these have been identified only by their genome, their DNA. More details can be seen in the 2016 web article by LA Hug and others and in "New Tree of Life Doesn't Look as You'd Imagine".

If amino acids were fabricated using mud and components in the early atmosphere and a variety of sugars were readily available (as we see from recent experiments they can be), it seems likely that short gene viruses can be made also. They can then join together, perhaps using the surface induced catalyst process, to form longer chains and eventually to form primitive cells, all the time exploring chemistry and seeking the easiest way to reproduce. Cells provided by far the best option. As soon as that stage and the cell division stage was reached, invasion of cells became the way to reproduce for small viruses. That road was much easier that the difficult evolutionary path to first form cells.

The transition from single cell life to multi-cell life is the next important transition in the evolution of life. Study of this transition may give us more clues about the origin of early life. As reported in 2015 in the on-line accessible journal *elife,* all it may have taken was one mutation more than 600 million years ago, to start the development from single to multiple cells. This Oregon based team studies cells called choanoflagellates. These cells exist today as sea sponges, both as colonies of independent single cells with tails for moving about and as a group of multi-cells. Their present day DNA structures are known, but in order to explore and understand the evolution from single to multi-cell cells that cooperate, the team needed to compare DNA structures at the time of the transition to see any different. They used a computer model to trace the choanoflagellates structures through **their evolutionary** history back in time. This required knowledge of more than 40 other DNA structures. These results point to a possible path for this transition. What is needed to make this a strong story, is to modify a single cell to be like that predicted by the computer, make sure it can divide, splice in the gene predicted to cause the transition and see what happens. Easier said than done.

As reported in *Scientific American, March 2016,* a study published in *Nature* by Stiel and others report single cell to single cell communication in bacteria via electrical signals. This allows corporative behavior to be exhibited by groups of single cells, and could allow new insight into details of the transition from single to multi-cell life.

There are still important events in the evolution of single cell bacteria that occurred before multi-cells came along. One of the critical ones is that early cells or proto-cells developed an amazing ability to degrade organic compounds and were able to drastically explore more chemistry. Early structures reproduce asexually; they just split the DNA down the amino acid structure to get two strands of RNA. Each then quickly makes two new DNA structures using the amino acids, sugar, and other molecules in the cell. Finally, the original cell divides to get two new cells. That duplication process is rather straight forward and chemistry tells us how. An enzyme splits the DNA structure, making two strands of RNA, builds up two new structures of DNA, and the cell splits into two new cells. There are some intermediate steps but these are well understood.

What other present day discoveries can inform us about the origins of life? An important technology has recently been developed. We have learned to change the coding of life, almost at will, and hence to create new life. It is called genetically modified life. As noted previously, the double helix structure of DNA can be thought of as a rope ladder with wooden sticks connecting the ropes, which are then twisted into a helix. The wooden sticks connect amino acids with a pretty weak bond called a covalent bond. In humans, DNA is separated into 23 pairs of long pieces joined together. These are the chromosomes. Twenty-three chromosome strands join together to form Sapien DNA which is about 2 meters long stretched out and about two nanometers wide. It is folded over to fit into a small 10 micrometers round cell nucleus. But the folding in not random as in stuffing socks into a drawer. An activator molecule of the DNA has to be physically near a gene to activate it and cause it to make those proteins needed for the task.

In the lab we have learned how to amplify small traces of DNA and that procedure had allowed detailed studies of vital processes. The amplification technique has the name PCR. Amplifying strands of DNA by the technology PCR (polymerase chain reaction) was developed in 1983 and is now extensively used to exponentially amplify small amounts of DNA pieces for study. The key word concept here is exponential growth. The process is well understood.

A number of these small strands of DNA, called genes, have been identified in all forms of life and isolated. Molecules can be assembled in the lab that target almost any desired gene (a sequence of amino acid chains). To these molecules biologists can add an enzyme that cuts the rope ladder in the desired location and replace it with the new gene. Something new gets created by design. I would be hesitant to call it intelligent design just because Sapiens is the name we gave ourselves.

The popular molecule today that targets genes is called Crispr and the enzyme that dissolves the peptide bonds connected the rope strands, Cas9. In fact, you can go on line and order a Crispr to target almost any gene in any life form (it works on DNA from all life forms), add it to your shopping cart, and purchase them for about $50 per target gene. Today it is possible to get into the gene splicing business for a few thousand dollars and by taking a few undergraduate chemistry and biology courses. This development offers great promise and also great dangers. Crispr and Cas9 molecules were first found in both bacteria, archaea, and presumably are in some of the proto-bacteria. They have been doing their job for a long time. They can be compared to text editing tools. We have only recently found them.

9
History of Life

By "life" we mean a thing that can nourish itself and grow and decay
-Aristotle

Bacteria and archaea life are thought to have started 3.5 to 4 BYA. About 2.5 BYA, a form of bacteria, still single celled, developed photosynthesis. It was a very efficient process and soon became stable in the environment. Slowly this event caused the external atmosphere to change from one dominated by carbon dioxide and sulfur to an oxygen rich one. Historically, it is believed algae were the first photosynthesis cell form to get out of the water and appear on land. Algae fed on the plentiful carbon dioxide; oxygen was its waste product. Photosynthesis life flourished. Our natural gas and crude oil are the products of these ancient single cell algae. The waste product of "plant bacteria" is the fuel for the other bacteria that fed on oxygen. Interestingly enough, oxygen is needed to attack metals in rocks, to create metal oxides that can then join together to produce the rich variety of minerals found today. This was noted by the geologists who created amino acids from a sulfur and a mud-water environment.

Soon bacteria, feeding on oxygen, formed a synergistic relation with the "plant bacteria". Slowly the earth's atmosphere and oceans changed to one of increasing oxygen content. Evolution continued and the first complex single cells (with internal compartments that specialized in tasks and called eukaryotes) appeared. These are found today fueled by oxygen but incorporating aspects of both forms of early life. In May 2015 the journal Nature reported a study that identified a new form of archaea that was called the "missing link" between simple archaea and eukaryotes. All life, including human life, today contains DNA components of archaea as well as components of bacteria.

Eukaryotes are like the early bacteria and archaea cells but have membranes surrounding the nucleus and other internal regions, an inner wall offering further protection. Most have mitochondrion (mDNA), a double membrane organelle inside the cell that has its own DNA (similar to bacteria). Most eukaryotes display rapid chemistry. I envision it was kind of like going to Home Depot and getting a new set of tools every week just to try them out, and then keeping the structures that work well in your environment. There is a class of single cell Eukaryotes called Dinoflagellates that is found today. This single cell displays photosynthesis and can take in stuff to try out more chemistry. Marine plankton, at the bottom of the ocean food chain, are members of the Dinoflagellate class. They show about 20 % archaea DNA

The structure of the Tree of Life has changed drastically since I left school in the 1960s. Supergroups, based on successful trials of Eukaryotes, rather than kingdoms are now the accepted starting classifications.

It is from eukaryotes joining synergistically together that all multi-cell life, starting with sponges and corals, developed. This began an important event in the evolutional path to me. Diagrams now show about seven major Supergroups of Eukaryotes with a path to present day life. All early life still had to hide under water to be safe from the ultraviolet radiation of the sun. These UV photons destroy amino acid structures and all life had to await the formation of the ozone layer to absorb out this ultraviolet radiation. Fortunately, the oxygen-ozone cycle triggered by the ultraviolet radiation from the sun is better than 98% effective in blocking the hard UV. After this development, fungi appeared on land (we are in that Supergroup), followed by plants less than 1 BYA.

The time between 800 and 630 MYA showed great variations in the environment, ranging from "snowball earth times" with oceans frozen over to greenhouse warming events. The atmosphere was high in oxygen levels and plants grew to giant size.

On the average, life was doing OK, with some growth and some extinction. Some eukaryotes had evolved to use calcium and started to develop hard shells and spines. But it was a relatively slow process. The fossil record shows predator- prey relationships which drives evolution faster. Then there was a huge mass extinction event, the Ediacaran Extinction 540 MYA. Over 50% of all life was wiped out. During the Ediacaran era most life had yet to develop skeletons or hard shells; organisms were mostly some form of worms or sponges. These left only a small trace in the fossil record. Only if they died in soft earth, that suddenly hardened when they decomposed could they leave an imprint and give us evidence of their life. Much of the life in this era appears today as fossil fuels. The composition of this former life is now being recycled into the environment, driving change which in turn may lead to extinction of present life. Interesting how that works. It is sort of like the story of rocks helping to create life and then life helping create beautiful rocks.

Soon the high oxygen level, then in the atmosphere, dropped to about its present levels. Following the Ediacaran Extinction, different life blossomed; it was the start of the Cambrian period. The amazing fossil records of these times, abundantly visible in the Burgess Shale in nearby Canada, give us detailed records of both soft and some hard shell life. Since then we have fossil records of 5 major life extinction events, 440 MYA, 360MYA, 251 MYA, 205 MYA, and 65 MYA. Dinosaurs dominated the land between these last two mass extinctions.

Following each new mass extinction, different, new life rapidly populated the environment. Then, only recently, came a number of humans and finally us Sapiens. There have been many biochemical experiments in the road to becoming me, many "false" starts.

The later chapters in the evolutionary story of humans has become more settled due mainly to advances in biochemical tools, specifically in extracting, amplifying, and analyzing DNA. These advances, in combination with dating of archeological discoveries, have led to a strong story of human evolution and movement around the earth.

Recently a team led by William F. Martin of Heinrich Heine University in Düsseldorf, Germany, reported progress on the question of the origin of life. They suggest a specific picture of the ancestor of all living things. LUCA, Last Universal Common Ancestor, is its name. It is estimated to have lived about four billion years ago, when the earth was only 500 million years old.

I quote from the recent NY Times article by N Wade:

"Their starting point was the known protein-coding genes of bacteria and archaea. Some six million such genes have accumulated over the last 20 years in DNA databanks as scientists with the new decoding machines have deposited gene sequences from thousands of microbes.

Genes that do the same thing in a human and a mouse are generally related by common descent from an ancestral gene in the first mammal. So by comparing their sequence of DNA letters, genes can be arranged in evolutionary family trees, a property that enabled Dr.

Martin and his colleagues to assign the six million genes to a much smaller number of gene families. Of these, only 355 met their criteria for having probably originated in Luca, the joint ancestor of bacteria and archaea.

Genes are adapted to an organism's environment. So Dr. Martin hoped that by pinpointing the genes likely to have been present in Luca, he would also get a glimpse of where and how Luca lived. "I was flabbergasted at the result, I couldn't believe it," he said."

The 355 genes pointed quite precisely to an organism that lived in the conditions found in deep sea vents, the gassy, metal-laden, intensely hot plumes caused by seawater interacting with magma erupting through the ocean floor.

Deep sea vents are surrounded by exotic life-forms and, with their extreme chemistry, have long seemed places where life might have originated. The 355 genes ascribable to Luca include some that metabolize hydrogen as a source of energy as well as a gene for an enzyme called reverse gyrase, found only in microbes that live at extremely high temperatures". A recent discussion on this (Earth Sky News, Aug. 28, 2016) gives information on this view.

So the mud layer, water, high pressure environment story, mentioned earlier, gets more supporting data and hence stronger. Plus, it pushes the origin of life as occurring shortly after earth was formed but after water appeared. Volcanic activity could have pushed sulfur using organisms up into Darwin's warm pool, where they evolved to using oxygen instead of sulfur. Indeed, as noted in Chapter 6, archaea is found today on the surface of the earth, even archaea using oxygen as an energy source is found. As usual more data and study is needed and the computer methods used to arrive at this story carefully repeated and critiqued; that's how science works.

10

Killer Viruses

Virus is a Latin word often used by doctors to mean, "Your guess is as good as mine"

Let's look at some of the ways viruses work today. Sapiens have discovered much about our immune system and how that system protects our bacteria cells from attacks by viruses. Unfortunately, there are huge numbers of bacteria and viruses; the good news is that most do not kill us. The bad news is that some will. Bacteria and viruses operate differently.

As we know, bacteria are cells with no confined nucleus; they contain small strands of DNA and RNA, along with their energy source and the proteins needed to operate independently, all encased in a cell. They reproduce by cell division. They are self-contained, so to get rid of them we need to kill them. Their cell wall contains a molecule which strengthens the cell wall. Human cells, those with a nucleus, do not have this molecule. Penicillin, the first anti-bacterial drug, inhibits this molecule in the bacteria cell wall from doing its job. As a result, the bacteria cell weakens, expands until it bursts and the cell dies. But we kill many bacteria, not just the bad actors with this method. Often, when the war is over, we need to take pro-biotics to put some of the good guys back in us faster than with the normal bacteria cell division process. We still have a residue of bad cells left and rely on our immune system to keep their numbers low.

However, penicillin, first discovered in 1928 and first used in humans in 1942, and most modifications of it are now ineffective. Bacteria, have evolved a work-around. New methods are needed to help our immune system. The emphasis is in development of molecules that get into the cell, and attack some molecules needed in intermediate steps for the production of proteins and cell duplication. Molecules like tetracycline have been developed which pass easily through the cell walls of both humans and bacteria. They are designed to attack the enzyme that uncurls the DNA of the particular chromosome that is responsible for starting the cell-division process. So the bacterial cell does not divide and eventually is passed out of the system. Fortunately, tetracycline does not do the same in human cells. Expect invading bacteria to evolve to develop another work around and also render these attempts ineffective. **Another anti-bacterial drug** will need to be developed if possible. Inside all life there are a number of wars going on daily; as we age, we lose more and more of these internal battles. Our immune system helps in this fight but that also weakens as we age. It's a tough world in there.

Infectious viruses operate differently and require a different protection plan. Globally, we had a recent experience with the very deadly (to humans) Ebola virus, and presently we have the Zika virus which is thought to cause birth defects. I start by describing the often deadly influenza virus.

Virus infections are pretty similar in operation. Their outcomes are not. There are 3 types of flu viruses: A, B, and C; influenza A is the one that infects humans. All flu viruses have a single strand of RNA made up of 8 segments, just eight amino acid pairs and their accompanying molecules wrapped in a cell covering, and two ends. This RNA virus manufactures 11 different proteins; two of which are responsible for the make-up of two different types of spikes on the surface. One, labeled H is necessary to get the virus through the human cell membrane into the cell and nucleus, the other, labeled N, gets the virus out. The composition of these spikes can change due to mutations when the cell reproduces. So, you have H_1N_1, H_1N_5 and so forth. All these can be "seen" by the immune system and their specialized army of scouts continually roaming about the body. They test the nucleic acid sequences on the ends of the virus, if a fit is found, the virus sticks, enzymes are released, and they chew up and break apart the virus. If there is no fit, they do not stick and the virus is free to drill into the cell and nucleus. Your immune system keeps a small number of each specialized guys for each bad virus you have had. They are able to reproduce very rapidly when called. Some members of this specialized army hang around for your lifetime but sometimes they need to be refreshed. The goal is to kill (dissolve is the kind word) the virus before it gets into the cell. When a type A virus gets into the nucleus it can splice itself into your DNA causing itself to reproduce and also to produce proteins, proteins that you may not like. Both leave the cell quickly and get about their jobs.

Viruses can exist a long time outside a cell. The flu virus in no exception. Environmental conditions, mainly pressure and temperature changes can break them apart and destroy them and some use this process to talk of a virus lifetime. RNA viruses have a very high rate of mutations. In the flu virus, mutations occur mainly on the two ends. They are labeled by different numbers on the H and N ends. Subspecies labels follow that by letters like H_{1a}

These ends have main species, 1, 2, 3, etc and also subspecies a, b, c, etc. A subspecies of H_1N_1 killed my grandfather Kearney. It was not fully identified until 2005. Only after 2009 did flu vaccinations become readily available for all throughout the US.

Vaccinations consist of injecting small bits of the known virus into your system so as to "school" or refresh your immune system to resist this cell invader. If the known flu virus tries to invade your cells your immune system can then identify them, build us an army (since it remembers how to do that), and destroy them quickly before they get into your cell. Once the virus gets into the cell it can take over the machinery, reproduce, make you very ill, and often kill you. The virus evolves, often within a year, with new in and out ends (H and N) that your immune system does not recognize. So you need a yearly "flu shot" and still you are susceptible to any new strains that have evolved over that year. If left unchecked the flu virus can kill the human host in a year or two.

Ebola and the Zika viruses are examples of very dangerous viruses that spend time in other life, mutate at different rates, and then re-infect humans. There are many in this class. They are called zoonotic viruses and the process called spillover. They are thought by many to become the cause of the next pandemic. Some virus can mutate to become airborne, others are spread by bats, birds and common insects like mosquitoes. These animals are called vectors and mosquitoes have been identified as the insect in which Zeta mutates and is spread.

When the Zika virus is injected in the blood of humans and if it is then passed on to a baby who is just developing an immune system, bad things result. Often it results in diminished brain connections and a small size of the baby head. The World Health Organization is on top of this threat but there is really no quick fix. Vaccines seem to be several years off. Major drug companies are reluctant to devote time and monies to this development; they have made major commitments in the past to developing vaccines for other viruses, only to lose a lot of money when the threat died out by itself and countries never paid them. The virus is spreading around the world. Other than warning women to avoid environments that harbor the insects or encouraging them to not have sex and become pregnant, the other focus is on removing the virus from mosquitoes. We cannot kill all the mosquitoes, but here is probably the best scheme to do this.

First, identify the bad acting sequence of DNA or RNA in the mosquito and verify it is only on one side of the double strand (viruses are usually RNA viruses). The Zika like virus has been known for a number of years now and I suspect this step has been completed. Next, use gene splicing techniques to remove the responsible (bad) DNA sequence from the insect's DNA, replacing it with another sequence if necessary so you "cure" the insect but do not prevent it from reproducing or dying. When these "fixed" insects get together with an infected mate and reproduce there are now three strands of good RNA and only 1 of the bad so the odds when two get together to form a new life are three to one of producing "good" insects. After a number of generations, the bad guys get wiped out. This is referred to as creating a gene drive. The final process requires just releasing some insects, now virus free, in a nice swampy area where they like to reproduce and then waiting for genetics to do its thing. I have heard it will take two or three years at best. Zika is truly scary.

So, how does all our present knowledge of biochemical activity give insight into our question, how did life start? The story implied by all the data so far is that there is a continuum between creation of non-life and life, and this progression is driven by natural processes. It is a strong story. There are gaps in this story but present studies are filing in many of these gaps. This gets us into the next section, the story of consciousness and how humans construct their beliefs and put together their truths.

IV. Consciousness

Consciousness, there are about 20,000 papers on consciousness with no consensus. Nowhere in history have so many people devoted so much time to produce so little.
Michio Kaku

We cannot define anything precisely! If we attempt to, we get into that paralysis of thought that comes to philosophers, who sit opposite each other, one saying to the other,' You don't know what you are talking about!' The second one says 'What do you mean by know? What do you mean by talking? What do you mean by you", and so on.
-Richard Feynman

The scientific and philosophical consensus is that there is no nonphysical soul or ego, or at least no evidence for that

-David Chalmers

11

Consciousness

I have been studying the straits and dispositions of the "lower animals" (so called) and contrasting them with the traits and dispositions of man. I find the results humiliating to me.

-Mark Twain

Louie Armstrong is quoted as saying *If you have to ask what jazz is, you'll never know.* It seems to many, the same can be said of consciousness. Consciousness is difficult to define. It is sometimes taken to be synonymous with "self-consciousness" or self-awareness. But we can be conscious of many things other than oneself (other people, the external world, etc.). The term "consciousness" is also commonly used to refer to a state of wakefulness. Being awake or asleep or in some other state such as coma clearly influences what one can be conscious of. When we became able to observe brain activity, and if none was observed, this became known as a brain dead state, quite different from a sleep or dream state where a lot of brain activities can be detected, but almost none that we can recall. Consciousness may also refer to the ability to have knowledge of or be conscious of something. And then of course we need to ask if other animals have a conscience, presumably a source of one's consciousness. Whatever it is, consciousness seems to involve:

(1) a structured brain, specifically in mammals the cerebral cortex;

(2) an awareness of self, other objects, and life;

(3) memories.

I limit the discussion here to the assumption that we need biological life to have a conscious; no thoughts here about the living earth or universe. As anyone who has been "put to sleep" during an operation knows we do not need to display consciousness all the time to have a conscious. We also know many life forms have a well-defined brain structure and display memory. So how about number 2? Again many life forms display awareness. So will this be a short chapter and we can conclude most all life has a conscious?

Alas, no. There is no agreement on what awareness means and consequently how to test for it. We need to try to separate intelligence and awareness. For me, intelligence is just the ability to learn. Sapiens display that all the time; so do other animals and even plants. It is perhaps easiest to observe in infant behavior when they are developing their first cognitive map. Some say that along with intelligence we need to look for displayed signs of connectedness, perhaps emotions, with other life and things in our environment. For the most part, displays of intelligence, emotions, and memories occur at the same time and are difficult to separate.

In 1994 an International Conference, titled "Toward a Science of Consciousness" was held at the University of Arizona. It has become an annual conference bringing together many disciplines. An April 2016 New York Academy of Sciences conference had a featured panel discussion "The Rise of Consciousness". You can view the discussion at New York Academy of Sciences Physics S4. It seems that the research community no longer holds to Descartes's separation of mind and body beliefs, but some still talk of the easy and hard problem. The easy problems are those that show a clear pathway to be explained in terms of neural mechanisms. The hard problems refer to things associated with experiences. To me there seems to be no separation between the easy and hard problems. But the easy problems are really not very easy.

Can an entity be intelligent but not have a sentient life (that is have a conscience)? Of course they can. Robots do this today and the rise in robotics has led to much discussion on this question. An interesting science fiction story, The Hunter Captain, address just this question. It is a great read at the author Baker's web site (see references).

I suspect everyone would agree that a virus or a bacterium is not conscious but that we Sapien are. So how about the rest of life. For example, are apes conscious? How about Chimpanzees or Bonobos? Some will claim that only humans have a conscience, even though many animals seem to exhibit human like behavior, express feelings in a manner like humans, and act like humans. A recent opinion piece, 2016 in the NYTimes by Frans de Waal (see reference list), raises some interesting issues about the use of words, for example the use of mating behavior for animals instead of using sex. Animals don't have friends, they have others they bond with.

De Waal also points out the use of the word anthropomorphism (from the Greek meaning human form) was first used in the 5th century B.C. against Homer's poetry that described gods as though they were human. Today it is used to censure the use of human descriptors for animal behavior. I point out that in order to survive Sapiens, like all other life species, must eat other life whether it is plant or animal life. Somehow we believe that if that other life does not have a conscience, it's OK to kill them and just eat them. We satisfy our hunger, get nourishments and have no thoughts about our feelings about the life we just ate; after all, that thing we just ate did not have a conscious. But just because that thought may be comfortable to us does not make it true. We need to include all life in the question of intelligence and consciousness.

Intelligence is something that can be measured or estimated pretty well. Just counting the number of neurons in the brain is one way. I look only at mammals. If you want to include invertebrates, you could start by reading the interesting little book, *The Soul of an Octopus, A Surprising Exploration into the Wonder of Consciousness,* by Sy Montgomery, 2015.

All mammals have a cerebral cortex component of their brain. Humans are near the top of neuron count with 20 billion), just a bit below the pilot whale, a type of Dolphin, with 38 billion. Our cousins chimpanzees have about 6.2 billion and horses 1.2 billion neurons. Your cat has 0.3 billion and your dog 0.18 billion.

How about the ratio of brain size to body size? Do Sapiens lead? No, the winner there is a shrew, not humans. But we are close. The number of brain neurons as an indicator of intelligence correlates well with our personal list of intelligence (except maybe for not leading in either category). About half of US households have pets; most pet owners agree their pets express friendship and awareness of others. Many critics argue we should not attribute emotions to non-humans; it is difficult not to do so. "Unlikely Friendships" by Jennifer Holland, 2013, is a collection of stories of animals with nothing in common, who display warmth and trust independent of species, the mark of true friends. The data as well as our everyday observations strongly suggest intelligence is on a graduated scale, and humans are at or near the top of the intelligence scale right now. However artificial intelligence is catching up. The evidence strongly suggests the development of intelligence is linear and that all mammals have intelligence. Indeed, all life displays intelligence.

But consciousness is different. To look for consciousness, consider a working definition of it as the ability to have many experiences, to store them in memory and recall them, to build connections to other memories which are then connected to emotions and pain. In addition, there is the need to observe signs of awareness outside of the self. It is the process of "understanding" and as a result to predict events outside of the immediate present. All Sapiens can do this unless some of the tools or connections in our box of what makes us conscious is not functioning properly (lack of neurotransmitters, lack of oxygen to parts of our brain, etc.).

Almost all living species show these signs. I offer two examples, this time from the western bird family: the Raven and the Clark's Nutcracker. In the 2015 book, Marzluff and Angell describe an experiment. Crows are well-known for making loud noises. A lot of them pass by the University of Washington campus. These researchers put a Dick Chaney mask on a grad student, had him walk across the campus waiving a stick and shouting at the crows.

The crows responded in kind by making loud crow sounds. Thereafter when he just walked across campus, they would again make vocal complaints and even buzz him. One day the student put the mask on upside down and set off across the campus. The crows were observed to fly upside down and then verbally complain and dive bomb the Chaney character as before.

The Clark's Nutcracker is a gray robin sized member of the Jay Family. Seeds from limber pine trees are its primary food source. The limber Pine grows well between 7,000 and 8,000 feet on wet, cold, snowy slopes in the western US. In August the bird gathers seeds from the cones and stores the seeds in the ground with a preference for 3 seeds in each spot. They will need this food during winter. The burial places are often separated by a considerable distance. When the snow covers the land and the bird needs to get at the cash of seeds for food it must remember where they are. The birds do a pretty good job using their memory of the layout of boulders and trees, but miss a few locations. For that reason, limber pines are commonly seen with three trunks. The pine cones are far out on limber branches, evolved so as to discourage squirrels from harvesting the nuts, but a perfect spot for a small bird to sit while harvesting seeds.

In both cases these birds display an amazing memory, an ability to plan for the future, and decision making. That they are quite intelligent, most I suspect would agree. I also think they both show signs of consciousness. They protect each other and **select a single mate.** When you hear a flock of crows making loud noises, often it is to protect one of their tribe. To observe signs of consciousness in other animals, we must walk in their shoes and look carefully. Heinrich in *Mind of the Raven* does it well.

There are two approaches to study human consciousness. One starts with philosophy and is a top down approach; that is, since we think we have a conscious we start there. The other approach is from a neuro-biological base; from the bottom up where we start from neurons, synapses, etc., examine what the small parts do, slowly study more complex parts, and hopefully learn how the small parts contribute to the whole. The goal is to meet somewhere in that vast middle region; the cognitive sciences area, a field including almost any discipline you can think of. An informative book about the middle region is "My Stroke of Insight" by Jill Tayler, a highly respected brain scientist. She describes her personal experience with a massive stroke to her left hemisphere and her journey over an eight-year period to "full" recovery. This and her TED talk on the development of the brain from childhood, through the teen years to adulthood offer valuable insights into the cerebral cortex structure and development.

The bottom up approach has been very fruitful, but the top down not so much. In fact, E.O. Wilson comments: *"I don't believe it too harsh to say that the history of philosophy when boiled down consists mostly of failed models of the brain"*. I believe that is kind of harsh but sort of correct.

According to Wilson, few present day neuro-philosophers spend much time incorporating recent findings of neuro-science. Patricia Churchland is among the exceptions. You might enjoy her Bill Moyer interview (see references) and her next interview there with a Canadian TV guy.

Next is the choice of words. Are the words conscious, sentience, essence, and soul all the same? Can you have a conscience without having a soul? Does all life have a soul? Many of these words are used interchangeably. Is a conscience something natural or supernatural?

Here is what I know about the origin of these words. The word soul comes from old English. In the

Proto-Indo-European Celtic language (~3,500 BC) the word for soul is anamon. In later day Greek it is psyche; in Latin anima; in Hebrew neshama or possibly mepesh. In western culture it dates back to the 8th century BC, to the stories by Homer. In the 5th century Orpheus believed the essence (soul) was trapped in the body and needed to get out. Pythagoras had a similar view. Starting from Socrates, then his student Plato, then his student Aristotle (470-322 BC), the soul became divided into two parts, one material and one immortal. This view more or less lasted until the 13th Century AD when Thomas Aquinas broke the soul into three parts to square it with Christian teachings. Plants, trees, etc. had a living soul that died with them. Animals, in addition, had a sensitive soul that allowed them to move and hunt for food. Humans had another soul, the rational soul. To me this was sort of a soul for all needs. It told us what we could kill and eat and what we should not eat (humans); the rational soul allowed for a God; it gave humans an immortal soul and an afterlife. This view seemed not to be seriously challenged for four centuries, when it arose again at the time of the scientific revolution. Science and religions started to take different paths.

The Soul Machine (2015) by George Makari gives us a lively detailed history of the transition during the 15th and 16th centuries of replacing the soul by the mind. Today science and religions often describe different realities; much more separated than in the days of Thomas Aquinas. The two have basically different starting points. Science embraces uncertainty and at its base is probability. Religions are based on certainty and authority. Science happily modifies or replaces theories as new information becomes available. Religion "reinterpret" old texts to try to keep up-to-date, or just denies new information. Signs of religions seem to reflect the ability to think outside the present and is used by many to be a sign of consciousness. These issues will be explored in a later chapter.

There have been a number of experiments that show animals and most other life can laugh, have self-awareness, can think ahead, have memories, can problem solve, display emotions, and have most all other attributes normally held as displays of intelligence and consciousness. De Waal, in his 2016 book, *Are we smart enough to know how smart animals are?,* has a number of examples showing this. We have found that all life uses some means of communication whether it is between their own tribe or to another tribe. For example, flowers communicate to insects which serve as the transport necessary for the flowers to reproduce. With most animals we can directly observe communications within their tribe.

A number of programs have been undertaken for the purpose of teaching animals, particularly chimpanzees, human words and their meaning. These have been well-publicized but have had limited successes of advancing knowledge. However, I do not know of any success or even attempts to teach humans to communicate in say dolphin language. According to Table 1, would the dolphins then conclude that humans have less intelligence, and would some dolphins believe, no conscious? Perhaps some dolphins would say no; we dolphins need to live in the other's environment to see how evolution have shaped the others. How smart are they in their environment? Unfortunately, all species have a difficult time living in an environment other that the one they evolved in.

I can almost hear the call to unite in todays op-ed piece in the Dolphin Express; *Something has to be done about these humans. They are taking over more and more of our environment. I know sometimes they have small boats that provide waves that we can play in and sometimes tease them, but lately they have Hugh boats making so much noise that we cannot even talk to each other. Plus, if we get close, some have witnessed members of our tribe being cut to pieces by their attachments that churn up the waters. Yes, something has to be done.*

12

The Human Journey

Do not go where the path may lead, go where there is no path and leave a trail
-Ralph Waldo Emerson

There have been many types of humans; now there is only one. The earliest discovered *Homo* species, Lucy, is thought to be 3 million years old. *Homo sapiens* have been traced back to Africa about 200,000 years ago and recently *Homo Neanderthal* to 176,000 years ago. In the short time span Sapiens and other humans lived in the same areas, about 40,000 years (~700 ALSs), Sapiens took over the world. All other humans went extinct. To see what short means it is helpful to look at this in terms of adult human life spans. Remember sixty years represents one human life span.

The split between chimpanzees, our closest relative on the tree of life, and *Homos humans*) happened about 100,000 ALSs ago. About 2 million years ago, humans we label *Homo erectus*, cut up raw meet with tools of stone. This allowed them to get more calories with less chewing. Faces and teeth became smaller, the changed vocal tract could allow more sounds to be produced, and increased calories could allow more brain growth. The spinal column was repositioned and this lead to improved walking, running, and communication abilities.

Table 1: Time-Line of Major Events

Years Ago	Adult Life Spans	Event
4,000,000,000	~100,000,000	Life on Earth started
3,200,000	~60,000	Early Homo species "Lucy"
750,000	~12,000	Homo Heidelbergenis migrate out of Africa
200,000	~3,300	Early Homo sapien in Ethiopia, Africa
170,000	~3,000	"Eve" Appears in Africa
70,000	~1,100	Homo sapiens Leave Africa
10,000	~167	Farming Communities
7,500	125	Black Sea Floods
2,000	33	Time of Jesus
60	1	Manned Space Exploration

Between one million and 500,000 years ago a of humans, *Homo Heidelbergenis,* migrated north out of Africa. They evolved to *Homo Nearderthal* in

Europe and later to *Homo Denisovan* in the east as well as into a number of *Homo* species throughout the world. About 70,000 to 100,000 years ago another large group from the folks who had remained in Africa and had evolved by then to *Homo sapiens*, started another migration north. They mated a bit everywhere they went with the earlier wave of humans, probably took the best genetics they could and became the sole humans left. We do not know why humans seem to migrate, but the evidence is that they did. One of the big reasons to travel is to explore life in a "better" environment. So a look at the environment in Africa may inform us about these migrations.

The Great East African Riff Valley, stretching all the way from the Mediterranean Sea down the Red Sea, past the equator to the Congo Basin, has many techonomic plates moving in different directions at different speeds. There are lots of historically active volcanoes in that area. There is evidence of quite rapid climate change from a dry arid place to one of lush vegetation and back, resulting in a swing from rainforests to savannas, about the times of human migrations. This could lead to a strong motivation to "get out of Dodge". So far this connection is just an idea, but for me a reasonable idea. Immigration could have followed animal food sources.

Cooking food will also provide more calories with less chewing; cooking began about 500,000 years ago, and perhaps became routine in the African group but slower in *Homo* Heidelbergenis. This could be responsible for the "advantage" Sapiens had over the natives when they first co-mingled. So far, this is another idea with only a few facts to support it. It is discussed in Nature, March 9, 2016. To fully integrate this idea into the story of Sapiens more data is needed.

Another idea, again not yet a story, starts with knowledge of the skills involved to make stone age weapons (Scientific American, April 2016), and how this development might have shaped our brains. We do know as our anatomy changed to adapt to the different environments in Africa and Europe, so also our manipulative abilities would change and this might have led to increased brainpower. So far no clear difference between Neanderthal and Sapien stone age tools has been found and this remains yet just another idea.

The amazing recent finds in a cave near the south tip of Africa and not as yet dated, may shed light on this idea. These *Homos* had a Sapien-like skeleton but a small brain cavity of ~600 cc. For comparison chimpanzees have a brain size of about 450 cc, and Sapiens 1,300 cc. There are hints that this find of human remains may fit in between the Lucy find and later humans.

Another very recent report tentatively dating Sapiens in southern China to 120,000 years ago could mean that migration out of Africa to the East started with a trickle earlier than 70,000 years ago.

The story of how *Homo sapiens* became the only humans on earth today, as well as the history of earlier humans and theories of why they became extinct; is a very exciting and active area of investigation today.

However, the big story, strongly supported by evidence, is that all humans originated in Africa and spread throughout the world at least twice, with the second wave (basically us *Homo sapiens*) replacing the humans that came first. These first humans where shaped by their separate environments, spread out North, East, and West, and developed into a variety of humans. Sapiens were able to successfully interbreed with these early humans. Parts of their DNA are found today in present day *S*apiens. For example, I have 1% Neanderthal DNA and most Sapiens of European origin have 1-3 percent. Africans, whose early ancestors presumably stayed there, have none. All present humans are Sapiens. It seems that everywhere we went we mated a bit with the ancient people who came out of Africa much earlier. There is evidence, based on differences in their physical structures that Sapien communication abilities ware superior to the others.

We Sapiens spread out from Africa ~70,000 years ago, only ~1,000 ALSs ago, and soon everywhere we went the record shows large animals and existing humans went extinct. Many believe this was the first global ethnic cleansing. It resulted in us being the only *homo* species left. How did that happen? One idea is that we just killed off the natives.

Could this idea be tested and become a story, or is it just an idea, perhaps even a wild guess? The gold standard is to repeat the experiment many times under the same conditions and look carefully at the outcomes. One may think this cannot be done. But let's outline an experiment.

Take a homogeneous group of very early humans living in Africa ~500,000 years ago. Send half to Europe where they are allowed to spread out, to go anywhere they chose, but to continue living in small isolated hunter-gather groups. Keep the other half in Africa with no constraints. By 70,000 years ago the many groups who traveled to Europe as well as the folks who stayed behind all have evolved into separate groups of humans, the Africans into Sapiens who got used to living in larger groups. Now, have Sapiens go out of Africa, intermingle with all the groups that came out in the first wave, and observe the results.

Realize that we just about have done that experiment; the results are in. In all cases, all the first wave of humans, now considered the natives, as well as all the large animals went extinct. It seems there is a causal relation between Sapiens arriving and extinctions. Could it be that we out-competed the natives for food and space resources in their home environment, we just killed them, or some combination of the two. We Sapiens invaded the home turf of the natives, so to beat then we needed an edge. Some believe it was our increased communication skills. The record is not yet clear, but it is strongly tilted toward the story of Sapiens just killing the natives and the large animals.

The story of being the "last Man standing" cannot be dismissed by simply saying the ancients went extinct. There is probable evidence that we Sapiens killed them. Extinction is just the polite way of saying that. There are pockets of our earth today where the "others", the different ones, those not in our tribe, are treated almost as harshly. In my lifetime a number of genocide wars have been initiated. The "others" are identified in a number of ways; by appearance, by social status, by possessions, and perhaps more commonly by religion.

Another idea by Houldcroft and Underdown, April 10, 2016, with the imposing title, *Neanderthal genomics suggests a pleistocene time frame for the first epidemiologic transition,* is that Sapiens had evolved to live in harmony with a number of viruses that proliferated on the African content (their immune system had evolved to kill off invading viruses as we do presently). But, neither Neanderthals nor other humans that came out of Africa earlier had developed these antibodies. Thence the idea is that he virus, transmitted by Sapiens, killed the others. To look for an example of this we recall the invasion of Europeans of the Americans and the introduction of smallpox to the native people. So we know this is possible but it is still in the category of an idea. Perhaps a number of things played a role.

We have a lot of information about our Neanderthal cousins. Like all things in the far past it is from the material remains that we gather this information. From those we deduce ancient cultures, religions, and beliefs. As usual this must be done carefully with due regard for the effects of the environment at that time. Fortunately, the environments are pretty well known. They include only the last two ice ages and are localized to the east Mediterranean area of our present world.

Neanderthal remains were first discovered in 1848 by British soldiers working in a cave in Gibraltar. In 1850, in the Neander Valley in northern Germany, bones were found in an old rock shelter. In the early 1900s a skeleton of a Neanderthal was discovered on a bed of flint rock chips. The skeleton was positioned reaching for a stone axe, also discovered. Soon many Neanderthals skeletons were found through Europe, positioned in what was described as a fetal position, as if coming in or out of the womb. Marsellin Boule, early in the 20th century unearthed a complete male skeleton in southern France, dressed him in clothes of the day, and exhibited him Thus was born the myth of Neanderthals as a dim witted ape-like creature. The belief was they did not have a conscience and hence it was OK to kill them.

That image of Neanderthals started to change about 1952 when Shanidar Cave in northers Iraq was discovered. Further explorations and advances in science over a ten-year period completely changed the view of Neanderthals. Today they are known to have a brain case somewhat larger than Sapiens, were a bit shorter than us, and had enormous strength. They were extremely agile and could run much faster than Sapiens. Even though my team may not have had advanced skills in passing and catching, if I had a team composed of Neanderthals, we would be overqualified to play in the present NFL.

A number of skeletons found in pockets of caves were not only laid on flint chips but were originally covered with flowers, identified by the remaining pollens, as being from medicinal plants. Most interpret this as clear signs that our Neanderthal cousins were spiritual, believed in an afterlife, could dream and wonder, and most importantly had a conscience. This is now much more that an idea; it is a strong story, almost a theory, with a lot of data to back it up. Neanderthals were much like we are today. Their brain cavity was larger than a Sapien's and they had a larger DNA structure than us.

All mammals have a cerebral brain. The development of a cerebral cortex is interesting. When mammals change environments from a protective womb to that of the outside world, they enter with barely sufficient knowledge to survive. At this time the cerebral cortex has a structure with many neurons but basically no connections to any memories; the baby actually needs to create memories. Humans are among the least prepared for their new environment. We observe this today with small children. Children with eyes wide open, staring with rapt attention at an event, are working hard; they are constructing their cognitive map and storing first memories. We Sapiens continue to do this our whole life. It takes a lot of energy and effort to put it together and maybe more to rewire the connections, to change our beliefs.

13

Humans Environments

It seems logical to us today that pretty things like flowers should be placed with the cherished dead, but to find flowers in a Neanderthal burial that took place about 60,000 years ago is another matter
-Ralph Solecki

It seems almost obvious that the environment can alter one's cognitive map. Our next excursion is to look at the environments of early humans and see how that may inform us. I know my environments have help shape my own cognitive map. The findings may also help to identify any signs of consciousness.

The temperatures of past environments that humans have experienced are displayed in the next page. Details of the various methods used to arrive at temperatures and times are readily available both on line and in more detail in the literature. This graph displays independently acquired data of temperature and CO_2.

Figure 1: Earth Temperature vs Time (NASA)

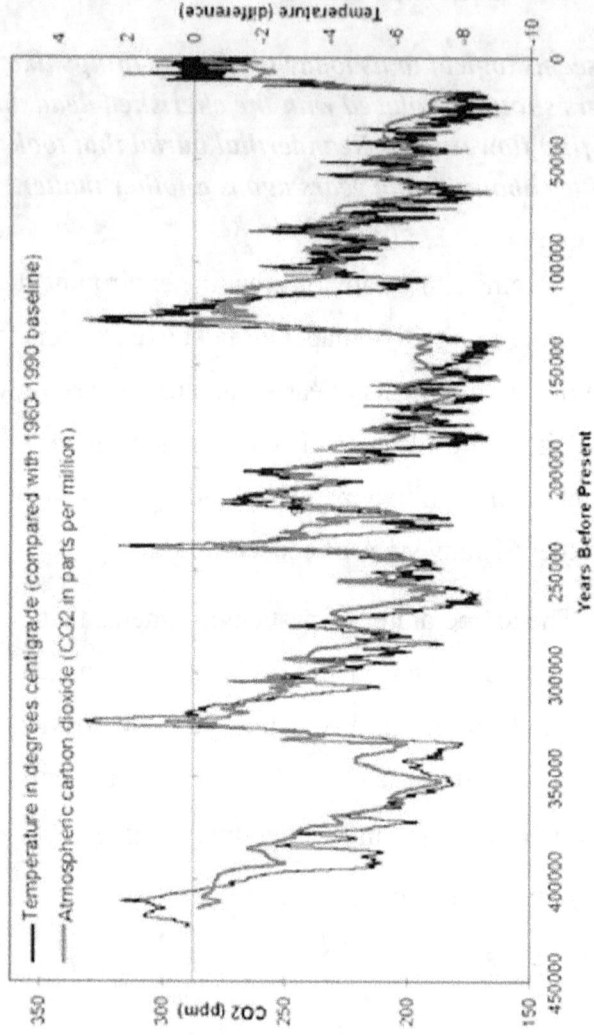

Our *homo* species is dated presently to about 200 thousand years ago (3,000 life spans ago); so this timeframe of this graph is related to the span of *homos*. Our tribe has survived two ice ages.

This NASA graph also shows the close historical correlation between temperature and CO_2. This graph shows older data. Presently the CO_2 levels are over 400 ppm and the temperature is rising. Humans are headed into uncharted environments.

To appreciate how short human existence on earth actually is, if the length of your arm represents the time of the earth's existence, a light rub with a file on your fingernail rubs out all of human history.

The first Neanderthals (and related Denisovans in Asia) appear in Eurasia 200,000 to about 400,000 years ago. As you can see from the temperature records, early humans experienced 4 very long cold periods when glaciers covered most of the land and sea levels dropped. In spite of this some found their way along the Asia coast to India, north to Siberia, across the ice bridge to Alaska, quickly populating the Americans down to the tip of South America. I find that amazing. Our cousins were great explorers.

Meanwhile, the folks in African who had not yet ventured north had developed into the first *Homo sapiens* (earliest remains identified is ~200,000 years old, in Ethiopia). There appears to be a significantly larger range of DNA among the present day African Sapiens than among the present day Eurasians. When the last ice age set in, some *Homo* Sapiens started moving north, peaking about 70,000 years ago. There are recent DNA data that suggests these Sapiens came from the area of southern Africa. Many settled where the Neanderthals lived, around the Black Sea. It was a glacial period. During this cold time the Bosporus, presently the small narrow shallow channel connecting the Black Sea to the Mediterranean, was above water. Passage was easy as was life in this warmer pocket of Earth. *Homo sapiens* apparently mated a bit with the "natives", multiplied, and soon set off on their own travels across the land on much the same path as earlier explorers. As noted earlier all the natives, as well as most of the large animals, became extinct.

How did Sapiens become the last humans standing? One idea is that before Sapiens migrated north they had undergone a "cognitive revolution" and had progressed much further than other *homos*. Here is another guess: perhaps driven by survival strategies and more advanced communication skills, the African humans who became Sapiens were then able to form larger cooperative social groups than other humans. That would give them huge advantages. We know the whole is often much more than the sum of its components. At that time, 70,000 years ago, food was readily available; we were just coming out of the last ice age and mostly all humans could use fire.

Until May 28, 2016, the earliest date of unambiguous human constructions was about 20,000 years old. Now with this recent dating of the findings in Bruniquel Cave in southcentral France, they are dated 176,000 years ago. Neanderthals were the only ones there at that time and the cave provides a look into their life.

The dating of the finds in the Bruniquel Cave to 176,000 years ago gives us not only the earliest date of humans in Europe but also provides a window on their culture. The cave itself is a recent discovery. In 1991 the cave was discovered by a caver and his son. According to The Atlantic article: *"Some 336 meters into the cave, the caver stumbled across something extraordinary—a vast chamber where several stalagmites had been deliberately broken. Most of the*

400 pieces had been arranged into two rings—a large one between 4 and 7 metres across, and a smaller one just 2 metres wide. Others had been propped up against these donuts. Yet others had been stacked into four piles. Traces of fire were everywhere, and there was a mass of burnt bones."

The cavers immediately notified archaeologist Francois Rouzaud. Using carbon-dating Rouzaud estimated that a burnt bear bone found within the chamber was at least 47,600 years old. But Rouzard suddenly died in 1991, and his work remained on the shelf until last year when a caver, Sophie Verheyden, who specialized in using Uranium isotope dating explored the cave and collected samples to obtain a more accurate date. The date of 176,000 years ago surprised everyone. We now know that Neanderthals made tools, used fire, made art, buried their dead, and most likely had language. They displayed advanced cognitive abilities. Except for appearances, most likely due to evolution in different environments, these humans were the same as early Sapiens. They were the same as us. A set of two images in the article by Lynda Pine shows how our view of Neanderthals has changed. Please open the web link given in the references and view the first two images in her 2014 article in Nauyilus. The second illustration imagines a Neanderthal family sitting in your living room today, probably watching TV.

Take a careful look in your mirror; catalogue present-day Sapien behavior. Presently we Sapiens are by far the largest predators on the planet, and we do not take old or young life as other predators do. We take out the adults in the prime of their reproductive life. For example, every year Sapiens take about 14 percent of the total weight of adults from the sea; ocean predators take about 1 percent. We destroy species at an alarming rate and, every once in a while, we destroy a bunch of other Sapiens also.

It is interesting to speculate on the size of stable groups that can form, be stable, and cooperate. Present studies show groups work well with a maximum of 50 or even 150 people, about the size of small corporative farms today. When more people are added it is necessary to adopt a structure, now-a-days called a government, with, hopefully an agreed-on common belief or purpose. I note that a common religion could supply a lot of this. Without a structure larger groups collapse. It was not until after 5100 years ago, 3100 BC, that we have data on the development of the first cities. Most likely the first city happened earlier but we have no record of that yet. This happened in the "fertile crescent", the land near Mesopotamia. This is only 83 adult life spans ago. With this development came the first government and taxes, those things that we all know and love. These people were called Samarians. Soon many city states all with different models but all with a government and a "religion" appear in the historical records. It is no surprise that we are still trying out different models of living together, still having wars, still using different religions to try to stabilize our particular group and grow.

Presently Sapiens seem to be on the same track of all other life, multiply, dominate, use all the resources available, and then go extinct. The big difference between Sapiens and other life is there does not appear to be any constraints on this Sapien Quest. Many believe we may be headed for a new catastrophe and when we go extinct most all life on the planet likely will also.

Since Sapiens became the single dominant human, there has been one notable local catastrophic event we have lived through, the flood near the Black Sea. I refer to it in the opening paragraph of this story as a myth, as a belief. Present day Plate Tectonic investigations have revealed a slip fault running across the Black Sea down the Arabian Peninsula. Movement along this type of fault results in the creation of a sink hole and as a result during the long last ice age the Black Sea was about 500 ft below the Mediterranean; the Black Sea was also isolated since the oceans were lower and the present day connection, the Bosporus, was above water. The waterlogged, silty soil of the Bosporus has been known to liquefy during an earthquake.

7,500 years ago the earth warmed and the oceans rose. The Black Sea again became connected to the Atlantic. It may have been a catastrophic flood with a giant waterfall of water filling up the Black Sea and wiping out people (this event is thought by many to give rise to the story of the Biblical Flood). But, Aku and his Sapien tribes survived as did Neanderthals in times before. The myth (idea) has become a story with some science to back it up.

Catastrophic floods have been well researched in the US Pacific Northwest and date between 13,000 and 15,000 thousand years ago. These Missoula Floods occurred multiple times. In these cases an ice dam on the Clark Fork River built up a large glacial lake covering present day Missoula. When the dam broke, as it did a number of times in this period, it sent a torrent of water down the river. This backed up all downstream tributaries like the Clearwater and Snake Rivers in Idaho and the Willamette in Oregon. Each flood carved and carried sediment to the flood plains of Oregon and Washington and to the Pacific Ocean at the mouth of the Columbia. The next time you taste the fine wines from the Willamette and Columbia Valleys give a toast to the Missoula Floods; they were responsible for depositing that fine soil. The floods were also responsible for the "underwater mountains" near the ocean entry point of the Columbia River, making passing over that bar to the deep water quite dangerous.

Flood myths are numerous among many cultures. They all have basically the same story. God punishes humans by sending a great flood. The faithful are saved either by a warning or by having boats available. I think almost all can be related to actual floods and the stories are passed down through many, many generations.

14

Brains and the Mind

The mind is just the brain doing its job
-Simon LeVay, 1993

The human brain is a very complicated organ. Here I give my simple overview of some brain processes. I rely on many sources and studies. In early civilizations, like those of Egypt, the cognitive and control center of humans was considered to be the heart. When a body was mummified the brains were sucked out and discarded but the heart and some other organs were left to be used in the afterlife. Now we know that it is the cognitive brain not the heart that makes humans different from other animals. But we also know it is a matter of degree. The goal of present day neuroscience is to understand how the brain works, and ultimately to understand how the brain and mind are related.

How do we study the human brain? Invasive experiments on living humans are allowed only in cases where there is evidence that the procedure will prolong life. There are no such universal restrictions with other animals, and it is from these studies and from studies of dead humans that much knowledge is gained.

Fortunately, non-invasive, or sometimes just a little invasive, studies are also allowed with the human's permission. It is not long ago that scientists needed to rob graves at night to get bodies to study. But today's medical students are placed into group and given a dead body to cut up and study and there is more technology and science to guide studies.

We know humans have a more primitive older brain similar to those in many other animals. It lies lower in our head, closer to our spinal column, where signals go in and out of the brain. In addition, humans have a highly developed component on the top, known as the cerebral cortex. All mammals have a cerebral cortex. It is more structured in human than in other mammals. The brain is folded over and stuffed into a protected hard shell.

For mammals, DNA has constructed the brain but individuals are in charge of making most of the connections in the cerebral cortex. All Sapiens share over 99.99 percent of the same DNA and about 99.4 percent of DNA with chimpanzees and a lesser extent with other mammals. Those small differences in DNA between Sapiens give rise to rather large differences in appearences. Genetic variations of Sapiens occur at many different scales; on average they differ by only about .002 percent; that small difference makes every human unique.

The human cerebral cortex, a double sheet of neurons about two millimeters thick covering the cortex, is considered the seat of the mind. It is about twice the thickness as that found in other animals. This, plus the increased surface area of human brains, allows more connections to specialized parts of the brain. Everything you can hear, feel, see, every thought you think, is the result of brain processes. These processes create pathways that link to the rest of our bodies.

Our brain is relatively small, about 1 percent of the whole body mass, but it takes about 20 percent of the blood supply. It is very active and consumes a lot of energy. Connections in the brain are accomplished by cells called neurons. According to the dictionary, "neurons are specialized, impulse-conducting cells consisting of the cell body and its processes, the axon and dendrites".

Dendrites get input signals from outside the neuron. There are a large number of them. The neuron body itself is small, about 0.1 mm in diameter. But, axons can vary in length from one mm to one meter.

The neurons have most of the features of other cells in the body but are larger and are imbedded in a fluid region. Neuron cells are polarized (one end is slightly positively charged and the opposite end slightly negatively charged) so that molecules can be passed down their length. This process then changes the voltage across the small fluid region, called the synaptic region, and causes a specific neurotransmitter to be released into the fluid space. The electric field that has also been set up in this process then guides the neurotransmitter across the gap separating neurons. Each of these specific neurotransmitters is structurally matched and coded to be accepted by cells across the gap and then directed to the proper region of the brain. Most of the time they are passed on to the dendrites of another axon, but not always. The 100 billion neurons in the human brain can each have thousands of synapses, and we humans have about 100 trillion synapses. They are mostly protected by a layer of mucus. Neurons either fire or do not fire so they are considered pretty simple. We obviously cannot have them all interconnected as then any input would activate our whole brain after getting an input signal and we would have no discrimination. We most likely need more than a few. But how many is a few? That can depend on the type and strength of the input signal and on the structure of your biological neural network, your cognitive map. Brain signals and the cognitive map are very complicated. Moreover, they probably are individual specific. There are chemicals that can inhibit or enhance certain pathways in the brain. Some are very helpful such as the use of an anesthesia during an operation; others can be deadly such as an overdose of any number of drugs.

Our brains are very busy. When we chew gum and walk at the same time there is a lot of stuff going on. A good whack on the head can often cause neurons to fire randomly and one experiences "seeing stars". Experiments have shown that when we get new information into the brain, new pathways in the synaptic regions grow; new pathways to the stored memory are made. Pathways are particularly strong if an event is associated with pain. So the old saying "beatings will go on until you believe what I say" has some basis.

All brain activity is accomplished by moving charges and science had developed a very nice non-invasive technology to track these connections. It relies on the fact that moving charges create a field. The technology is able to detect these fields and track them in real time. To learn more just do an online internet search for *MEG imaging*. A new map of the brain, detailing nearly 100 previously unknown regions has recently been published in the journal Nature by a research team at Washington University (see Glasser et al.). They are part of the Human Connectome Project which seeks to get a complete map of the human brain.

Most of the cells in our body die and get replaced every few months, but neurons do not. If they die, (usually from lack of oxygen or as a result of aging) the brain is often in trouble.

The only external input to our brain is from the five senses; through the eye, nose, ear, skin, and mouth (The Vulcan Mind-Meld still remains in the Star Trek fiction realm). Output, needed to communicate in societies, is limited to language, gestures, and writing.

Most of external information into our brain is through the eyes. There are ~120 million rods and ~6 million cones (in sets of 3 for color vision), but only about 1 million connections to the brain. Most of the information hitting our retina never makes it to the brain; priority is given to signals from photo-receptors that change. So what you see at any time is mostly made up by your mind. It depends on stored memories, even long ago stored memories, and beliefs. The time delay between light hitting your eye and some recognition of it by the brain varies between 13 to 100 milliseconds. There is always the time spent in pre-processing and then the travel time required to get to the brain. In this sense, you are aware only of past events, and remember in this time your brain makes up about 99 percent of that you "see". If the scene is not changing and you continue to look you can do much better. But your memory is not a good way to store data. The hole in the retina where there are no receptors, your blind spot filled with 'wires" leading to your brain, is an easily observed example of your creative brain at work. There are a number of on-line sites where you can explore illusions and your brains' creativity. The cerebellum at the base of your brain controls all this. About half of all the neurons in the brain are stuffed into this small cerebellum.

The considerable difference in eyewitness accounts by different people of the same event when each person is telling their version of the truth has long been recognized. This inaccuracy of eyewitness accounts is becoming increasingly recognized in legal proceedings.

It seems as we age the response speed of our information systems, including transmission to the brain and memory retrieval slows. This is most noticeable in my case in the sound system. When people talk too rapidly I simply cannot make sense of what they are saying, even when I know the words and their meaning well. Basically this results in nothing much getting stored in my brain for later use.

Our "imperfect" memory does however, get stored in our brain. It can be strengthened and connected to other memories for later recall by revisiting this memory and making more connections.

When we learn something, whether it results in improvement in activities such as tennis or just gaining any knowledge, a key question is how is it stored in the brain? Structurally changes in the brain's synaptic regions are involved; our brain develops more connections and many parts of the brain are linked. We know the physical regions of the brain where different memories are stored. Short term memory, languages and spatial memory, gets stored in places like the hippocampus, which also produce neuro-transmitters which are then used in the synaptic regions, and then makes molecules to dispose of the "old" used neurotransmitters.

Skills get stored in places like the cerebellum. Long term memories get stored in many locations often called association areas. When many connections are made to other storage areas, your memory gets strengthened. It has long been noted that facts are better remembered when associated with images or stories. We know that by constant repetition we can build up connections and train our brain. This may partially explain the difference between liberal brains, that embrace uncertainty and change, and conservative brains that need certainty, like the status quo, and resist change. It has been claimed by some that institutions of higher education are places where students are trained to be liberals. There may indeed be some truth in that claim. On the other hand, I believe Universities foster an environment where all are encouraged to question and examine their beliefs, to construct or modify their own beliefs and knowledge. There is developing data that indicates memory get refreshed every time it is recalled. So it really not like books stored on a shelf, pulled out, read, and then put back. The pulled out, read part is not the same as the put back part and your memory changes each time you recall it and then store it.

15
Myths and Religions

If God did not exist, it would be necessary to invent him -Voltare

Religion is an insult to human dignity. With or without it you would have good people doing good things and evil people doing evil things. But for good people to do evil things, that takes religion _Steven Weinberg

I suspect many know the myth of Adam and Eve from the book of Genesis. I know I learned about this and more in Sunday School where I had to be while my folks attended Sunday Mass. I learned that God created Adam and then, seeing Adam needed a companion, took Adam's rib and from it created Eve. Also, there is the tale of eating that apple. God allowed the devil to give the new couple the apple test. Unfortunately, both failed the test and man has sort of been screwed since. The devil is pretty tricky and gets us to do bad things sometimes. At first God favored the Jews and gave them 10 commandments so they would know how to behave. After a while they did not listen and so he started the Catholic religion. Now God is on our side.

I learned all that quickly and then got bored. I spent most of my time in Sunday School (we called it Catechism class) thinking about important stuff like sports and radio shows like "The Shadow". I stopped going to Sunday School when I started third grade. That fall I transferred from public school to MacDonald's K-8 school, the school associated with St Anne's Catholic church. I guess my folks decided my immortal soul needed more training. By that time, I could manage to sit quietly, but never still, for almost the entire 50 minute Catholic Mass period.

After grade school I spent 4 years at Bishop Bradly High, the Catholic High School and then college and graduate school. I learned a little more in high school about the catholic religion but not much more than I did in Sunday School. Although Kathy and I were married in a Catholic ceremony and I attended Mass regularly every Sunday until my mid-thirties, religion did not have much of an impact on my life. It was not until I was in my firth decade that I started to read more about different religions and to learn more about myths and how religions came about.

Here I describe some of the things I have learned. The Greeks had a different myth from the Bible; the story of Pandora, detailed in the 8th century stories of Homer. The Greek Goddess Pandora, the first women, was created by Zeus out of mud and water as a punishment to man. At that time Zeus was mad at man. Zeus gave Pandora a jar (in about the 16th century it became a box) with instructions not to open it. Pandora just could not resist, she cracked it open and released evil, coupled with a small amount of hope, to the world. No putting these things back in the box; Zeus won and humans had to live with it.

Both of these are myths but only one is "sort of" believed today. I think the Greek myth is a bit more fun that the Adam and Eve one of eating the forbidden apple and that rib thing.

How about earlier times? Is there any evidence on how religions started? To make headway on these questions realize that ancient myths, the oral stories verbally handed down in time and later written down, are not to be dismissed out of hand, but need to be examined carefully. They often describe actual events. For example, the Rock of Gibraltar, presently an English territory just off southern Spain, and its twin off the African coast are called the Pillars of Hercules. They form the gateway to the Atlantic Ocean from the Mediterranean Sea. In Greek myths Hercules splits the two continents apart and in another puts then together to keep the sea monsters out. There is robust evidence for this opening and closing of the strait from plate tectonics studies and the rise and fall of sea levels.

I found it interesting to speculate on the origins of religions. The most probable idea of many, including me, is that it happened gradually, perhaps in step with consciousness, during the long development period leading to the Cognitive Revolution. Perhaps even before the first migration north by the humans who were to become known as Neanderthals. All early humans were hunter gatherers and were dependent on animals for much of their food. They valued all life. In order to assure a good supply of food they had to please the supplier of various animals; their way of doing this was to offer sacrifices, sometimes even their own children, to the Gods, who were the suppliers of that food. Gods in the myths and stories of hunter gatherers most often appear in animal forms and had special powers. Eating the heart of a fresh kill was thought to incorporate that animal's strength into you and with that, part of the god represented by that animal as well. After the Agricultural Revolution and the domestication of some animals it became the custom to sacrifice some of your stock, like lambs, to the gods. Many of these myths and practices exist in present day religions. It seems that many started by promising life would be better by following their religion. When things did not get better, the theme changed to mean not in this life but in the next life. Most theist religions are really polytheist ones, whether the lesser gods are called saints or just lesser gods. When did the concept of the next life come about? What evidence do we have about early religions?

Here is what I learned. Our ancestors liked caves. Caves provided safety and warmth. Caves were their homes. Displays of religions with burial sites, signs of belief of an afterlife and art are considered by many as displays of consciousness. Caves are a good place for present day searchers to gather information on early humans.

The Bruniquel Cave, a 176,000 year-old Neanderthal site located in southwest France, is only starting to be fully explored. Already the artifacts are being interpreted as an indication of religious belief. In order to arrive at a full connection to religion we must await findings and further informed interpretation of on-going investigations. It is an exciting discovery.

We discovered, in the amazing finds at Shanidar Cave in northern Iraq, that early Neanderthal groups took special care of their dead members and often held dinners with the remains close by. That is believed by most as a strong indicator of religious beliefs and of consciousness. The Shanidar site is dated to 35,000 BC. Human Neanderthal remains have been found there resting on a bed of stone tools and covered with flowers. Careful DNA study shows these flowers were from medicinal plants. Perhaps early humans believed, as I do now, that a person dies twice. Once when all signs of life ceases and then a second time when no longer are his stories and thoughts recalled by his clan. Perhaps the Neanderthal's would be pleased to know that I am proud to have 1 percent of their DNA.

Altamira, in northern Spain discovered in 1878, dated to 16,000 years ago, and Lascaux, France (discovered 1940, dated 13,000 years ago) are detailed examples of cave art by Sapiens. These are the oldest artifacts, so far, that indicate very clearly humans' ability to look from outside the self. Many others artifacts have been discovered in a variety of sites indicating the same. These give rise to the story that human intelligence and consciousness evolved in tandem. I find it interesting to ask why.

Here is my take. Religions have their roots in humans needing to believe in cause and effect, partially in order to cope with the death of tribe members. All they could observe in their world had a cause and effect relation; it was natural to assume everything did. Slowly the idea of gods came into the world, some good, some bad; all modeled after observed human behavior. Early people grew to believe and live in both a material and a spiritual world. All humans seemed to display a need to learn about their world and to explore new things, to ask why. How has this lead to our present state? What do the major religions look like today?

I jump to 2,000 years ago when Jesus lived and the start of the Catholic Religion. Most are aware of this religion through the New Testament Bible stories. The Bible contains probably the most influential stories in western history and has been studied extensively by scholars. Its main character, Jesus of Nazareth, was a Jewish apocalyptic preacher who talked about God's kingdom soon coming to earth. By the word soon, he meant within the lifetime of those listening. He is thought to have been born about 4 BC (because Herod died then) and to die at about age 32, in Pilot's stint as governor from 26 to 36 AD. Jesus spent the last 3 years of his life as an apolitical Jewish preacher. He spoke Aramaic and his followers were uneducated people from around Galilee. At that time there were a number of start-up religions in the area as well as a number of prophets, many of them apocalyptic preachers, with messages similar to that of Jesus. For example, Apollonius of Tyana (an ancient city in southern Turkey) was a contemporary of Jesus and details of his life mirror that of Jesus. Apollonius was born in a miraculous manner, had disciples who thought he was the son of god, did miracles, and at death ascended into heaven. He was pagan and his teachings did not last much beyond his death; not so for Jesus.

    Most of the religions 2,000 years ago had many gods (polytheistic). It was quite common for these gods to become human and humans to become gods; that became part of people's belief systems. The Jews professed to have one god and their preachers were usually apocalyptical. These preachers were more or less tolerated by the Romans. Jesus was crucified by the Romans not because of his preaching but because he claimed he was to become King of the Jews on Earth and the Romans would not put up with that. They were in charge. According to the gospels, Jesus said when the Son of Man arrived, he Jesus, would be made king of the coming kingdom (on earth).

We know much about his life from the writings in the Bible, formally called The New Testament Bible. It consists of 27 books by 14 or 15 authors. All were written in Greek and were written by folks at least the generation after Jesus. The Bible has four Gospels describing the life of Jesus, the Book of Acts, 21 Letters written by Christian authors (13 by or attributed to Paul), and the Revelation of John, an apocalyptic view of the end of the world. Of the 4 Gospels three are very similar and are thought to be written about the year 85, exact pages appearing in each as if copied. These are called the synoptic (seen together) Gospels. John's gospel, written last about AD 95 is quite different. Only here is the mention of Jesus' divinity. Most scholars believe that the John of the Gospel is different from the John of the Revelation. These stories were written about five decades after Jesus died by people who most likely never knew Jesus. They were written by people who did not live in the area. They were written in Greek; a language Jesus did not speak.

The stories chosen for the Bible do not describe Jesus' childhood; the only mention is that Jesus left Jerusalem with his parents at age 13 (Luke 2). Then we find nothing about his life until he appears at age 29, preaches for 3 years and amazes folks who hear him speak (Mark 6, "How did he come by all this? What is the meaning of this wisdom that has been given him, and of all these wonderful works that are done by his hands? Is this the carpenter, the son of Mary?"). Did all the scribes and story-tellers have a 16 year "writers block"?

To get some insight to these "lost years" it may be helpful to look at the environment in and around Judea at the time of Jesus. Judea was an important center for trade between the West and India and it is known historically that Buddhism was a well-known religion in the area. Buddha (560 BC) was also believed to have performed miracles, healed the sick, walked on water, fed 500 men from a single basket of cakes, and taught many the same things Jesus did.

One idea about the "lost years" of Jesus' life is that he spent that time doing what was normal for a young Jewish boy at that time. He studied the Jewish Bible in the synagogue, worked the land, and spent some time studying the various religions in the area. Remember, the busy silk road trade route passed through the area and he was exposed to many different ideas. We will probably never know details of Jesus' "lost years" but we do know about the trade route and most agree on the story of how a typical young Jewish boy spent his days. So there is some soft evidence to support this story. Whatever he did, when Jesus presented himself to be baptized by John and enter public life, he wowed everyone.

There is little doubt that Jesus existed and was executed during the time of Tiberius by orders from Pontius Pilate. Tacitus (55-118 AD), the Roman historian who was no friend of Christians, describes these events. Tacitus wrote in Latin. Josephus in the first century AD also describes these events.

Today there are no originals of any of the early Christian texts. Everything had to be reproduced by hand, often from an incomplete copy and modified by stories passed through several generations. Until the 4th century these stories were written on papyrus; after that on parchment. There are today no complete copies of texts written earlier than the 4th century.

By the second century many gospels had been written; all different and all in competition to be read at services. There was a need to decide which ones should be included in the canon of the church, the Bible.

It took three centuries to settle on a unified belief about Jesus. Most believed he was god but was he god before and while he was man? Did he become god only after he died? Which of the many Christians writings should be believed? At that time Christianity was splintered. When Constantine was emperor of Rome he had a lot of problems keeping his empire together and he reasoned that if Christianity could be recognized as the only true religion, that would help his empire rally around one voice and may help keep it intact. He organized several "councils of the church". The first, the Council of Nicea, was to decide which writings should be deemed true and written down so "all would understand". He hoped this would unify different Christian beliefs. For him it was good politics. From such a shaky beginning, the New

Testament (the Bible) was put together in writings and proclaimed the Truth. Christianity was launched on the road to becoming the dominant religion of the West.

Predating Christianity was Judaism. God and Abraham (very loose date of about 2,000 BC) entered into a covenant. God said the Jews were his chosen people and would live in the promise land, presumably present day Israel. Their Torah, the first 5 books of the Hebrew Bible and the Catholic Old Testament Bible, was orally told to Moses (1393-1273 BC) on Mt Sinai when Moses was leading the Jews out of Egypt. It is thought to be handed down orally and first written down in about 550 BC.

After Christianity got started, came Islam. Islam, meaning "submission to the will of God", today has over 1.6 billion members. They are all called Muslims. It is second largest religion in the world next to Christianity with 2.2 billion members. Like most current religions, Islam plays an important role in the modern world. It has one God, Allah. Abraham, Moses, Jesus, and Mohammed are its prophets. According to the Qur'an, Islam has always existed. Mohammed (570-632 AD), at the age of 40 and living in Mecca was visited by the angel Gabriel while he was mediating in a cave on Mount Hira, near Mecca. He was told to just recite, he did; then the angel told him his words came directly from God. I understand Mohammed memorized the words and later wrote them down. Quite a good feat. This was repeated often over a 23-year period. The complete manuscript became the Holy Qur'an. By this time Mohammed had become a preacher. In 622 AD he was encouraged by the political establishment to leave and he went to Medina. (The Islam calendar starts that year). He returned to Mecca after almost 10 years and died there in 632, about 22 ALSs ago. During that period Mohammed became accepted by most all as the final prophet. He was also the political leader in the region. Tales of Mohammed's life and his beliefs, the religion of Islam, quickly spread east and west, mostly by war and success.

But this spread was not a quest to convert folks; it was about gathering more land. In this, religion was used as a justification for the political action of starting wars and killing the opposition. Indeed, the Qur'an has passages that justify this type of action. But then, the Catholic Bible was also looked on as a justification for the Crusades.

When Mohammed died in 632, there was disagreement on who should now lead. Some wanted the son-in-law of Mohammed, Ali, to be Caliph (meaning successor, the political and social leader). The other camp wanted "the most qualified" to be the successor. The split between the "blood-line successor" (later called Shia for Shiat Ali, the Party of Ali) and "the most qualified one (later called Sunni, People of the Tradition) started. The most qualified camp candidates were the first three successors; The fourth Caliph was Ali; he was later murdered along with his two sons and the split between Sunni and Shia widened. Violent wars continue today between these two factions and their various sub-tribes.

A kind description of the differences in these three religions is: Judaism holds that one becomes a descendant of Abraham through birth, and Christianity that one becomes a descendant through faith, but Islam holds that one becomes a descendant of Abraham through both birth and faith. These three religions are closely related Their canons reflect the different environments and common belief at the time they were written.

There are a number of other stories and myths, some first recorded in human memories and then later in print, that are known.

The rigidity of human belief systems is illustrated by a look at recent attempts to form "religions". This list follows that described in the 2009 book, Beyond Cosmic Dice" by Schweitzer and di-Sciara.

2000: In Uganda, Credonia Mwerinde convinced greater than 200 members of the Restoration of the Ten Commandments to die by setting fire to their church.

1990: Houston teen age Vernon Howell dropped out of high school, moved to the town of Waco Texas and changed his name to David Koresh. He believed he was the Messiah, appointed by God to rebuild the Temple and destroy Babylon. He was the reincarnation of both King Davis and King Cyrus of Persia. At least 131 of his Branch Davidians followers went to live with him in his compound. Many of the young women became pregnant by their Messiah. At least 80 members and 4 government folks died in a government led attempt to allow compound members to leave freely if they wished.

1997: Marshall Applewhite prophecies that the Hale-Bopp comet was the sign (at last!) for followers to shed their "containers" and 39 members of the Heaven's Gate cult took their lives so as to more easily get to the spaceship just behind the comet

1978: Rev. Jim Jones convinced 913 of his followers in Guyana (Jonestown in South America) to drink cyanide-spiked Koo-Aid and commit suicide. He claimed to be the divine reincarnation of both Buddha and Jesus. It was the largest mass suicide in modern history.

Early 1800: In upstate New York a boy made money by claiming he could divine the location of water and also of buried treasure. When he failed he was usually run out of town but just moved on to the next town. It did not help that he had a habit of "charming" the young women in the town and upsetting their parents. Finally, in 1827, after he failed again in Susquehanna Valley, New York, he married Emma Hale, the local girl of his choice and the first of his 40 wives. Joseph promised to change his life and work hard. Within a year his father-in-law discovered him sitting behind a curtain with a Bible, speaking long passages from it with Emma outside the curtain writing everything down. Joseph claimed an angel led him to a place where upon digging he found two golden plates (presumably nearby in upstate New York) written in "reformed Egyptian", along with instructions to translate the symbols into English. Hence, came into our world the Book of Mormon. No one but Joseph has ever seen these golden plates; he claimed instant death for anyone but him who even looked at them. They have "disappeared." Joseph Smith again was run out of town and went west along with some of his followers. Mormon followers settled in a nice secluded place near a big salt lake and large mountains, and the religion grew. By then Joseph Smith had died and Brigham Young had taken over. Tithing by its members and business investments today provide the church with the money to carry out their missions, one of which is to grow in numbers. The religion does not recommend the members commit suicide but does have an active missionary program.

    The book Ghost Dance by Weston LaBarre lists more stories associated with other religious beliefs.

    To classify followers of these religions as crazies or just nuts, misses the point. They are no more crazy than fervent Jews, Christians, Muslims, or members of other "main stream" religions today. The ones that recommended their followers commit suicide had little chance of longevity, but not so Mormonism. That religion is strong today.

Beliefs that seem to be cemented into one's cognitive map are quite resistant to modifications. These beliefs represent certainty and members are supported by their group.

Miracles are claimed by some religions to be observations that "scientifically" justify the beliefs of that particular religion, usually their version of God and their particular rules. There are many "holy sites" where miracles occur. Muslims gather in Mecca, Saudi Arabia, Hindus go to the Ganges River in India, Christians go to Lourdes and Fatima. I know a little about the last two sites that have miracles associated with them.

I have visited Fatima, Portugal. Fatima is an Arabic name; Mohamed's daughter was named Fatima. Now it is a Christian site. In 1917 three young girls claimed to have several visions of Mary, the mother of God The message seems to be "If My requests are not granted, Russia will spread its errors throughout the world, raising up wars and persecutions against the Church. The good will be martyred, the Holy Father will suffer much and various nations will be annihilated." The main request was for Russia to become a Christian nation. Of course, WW I was on-going at that time and that naturally dominated the environment. I have watched many believers traveling on their knees the mile-long path of the Stations of the Cross, followed by praying at the spot where Mary is said to have appeared. True believers, hoping they are good enough to be granted a miracle, are abundant.

As far as I know only Lourdes among all the holy sites, claims to have scientifically validated miracles, and there only medical miracles are considered. In 1858 at Lourdes in Southern France, a 14-year-old girl claimed to have several visions of Mary and then a spring arose to make the holy baths where miracles are believed to happen. Presumably a group of physicians are on call to evaluate every miracle claim. Medical records before and after the miracle are required and collected. If there is no known medical explanation the claim is passed on to the International Medical Committee of Lourdes and if approved sent to the claimants local Bishop. As of 2014, since 1858, 7,000 people have reported to be cured by a miracle and their claim submitted; 69 claims have made their way through the three step process and have been classified as miracles. As described by Jo Marchant in her 2016 book Cure, even if the miracle is found later to have a medical explanation, it can be said "a miracle is just an interpretation; a Bishop believes that a person has received a gift from God". Miracles are very difficult to prove except to those who already believe in them.

Then, there is the well-known placebo effect. In the medical area this refers to the fact that people seem to recover after receiving fake medicine. I know this effect well after participating in several phase 2 and 3 clinical trials to evaluate new medicines. It is a real effect and raises the question how can a simple belief have the power to heal? We can put this in a question that can be tested: Can the conscious brain exhibit some control over the immune system? Evidence so far gives a weak story that says yes, but details are sparse. It is an emerging area of research. I believe this weak story will be strengthened as humans continue to evolve and explore. Future generations will write of this placebo history.

Today, often with politics thrown into the mix, the split between science and religion is increasing.

Science offers a very different view of reality than religions; but its strides are rapid, profound, and at times somewhat disturbing. In 1959 C.P. Snow gave a now-famous talk to the Cambridge University community titled "The Two Cultures and the Scientific Revolution". Snow talked about the disconnect between science and humanities. He encouraged folks in both fields to build bridges, to close this gap so as to allow a better society to evolve. I believe Snow should have focused on the disconnect between science and present day religions. There is no conflict between the sciences and humanities; both are creative, both ask not only why, but why not. However, there is an increasing gap between science and religion; to the extent that many people today do not believe in science; do not believe in observations and data.

The differences between science and religion are large. An idea, or hypothesis, which cannot be tested and has no basis in previous observations remains just an idea. It is faith only if one believes the idea to be fact. Science has the ability to be changed and to grow in scope. Religions based on faith are almost the opposite. The canons of most all religions are written down, usually a necessary step so all will know their truth and how to act. These faith-based religions are particularly resistant to change. They have had a difficult time trying to do so. Science is based on observations and logic; religions on sources of authority.

Some will argue that one needs religion, or at least a theist belief, to develop a moral code in our cognitive map, a code that leads to a model of how we want to interact with all our Sapien tribes and how to live that code. I think we need science to develop our moral code. Religions have not done a great job.

It seems to me there are four choices:

1) Keep the code you learned as a child, either by having someone tell you or being told by your environment.

2) Look for an existing code after you reach the "age of reason"

3) Develop your own code after you reach "the age of reason"

4) Some combination of the first three

All these choices are dependent on the environment. Children learn by copying; your initial moral code and cognitive map is made up in childhood.

If we take religion as an indicator of our moral code, choice number one is pretty popular. A 2015 Pew study found 25% of US population belong to no organized religion, up from 16% in 2007. The study concludes "A growing share of Americans are religiously unaffiliated, including some who self-identify as atheists or agnostics as well as many who describe their religion as "nothing in particular". I have observed many atheists have a much stronger moral code that a number of "religious" people I know.

16

The Future

The highest activity a human being can attain is learning for understanding, because to understand is to be free
-Baruch Spinoza

We are star stuff which has taken its destiny into its own hands
- Carl Sagan, Cosmos

We all need to take great interest in the future because we will spend the rest of our life there.
-Unknown

If you don't know where you are going, you might not get there
-Yogi Berra

Many of us seem to dislike any uncertainty. Many want an authority; someone or something to assure them, to show them truth. However, science cannot prove anything; never has, never will. If poof is needed, mathematics is the place to look. Science can provide evidence based stories that can be used as a guide to predict. Nothing can predict the future with certainty. The best we can do is predict various futures and give probabilities for each based on past knowledge.

All the evidence I know of, leads to the realization that we Sapiens have free will and can choose how our cognitive map (basically our belief system) is hooked up. If we don't like it, we can change it. As children our belief system is largely structured by our environment and our own observations. We gradually take control and are considered to be responsible and fully in control when we become adults, about 18-20 years old. It's a lot easier to build up our belief system in our formative years as well as to develop skills such as speaking and understanding a foreign language, than to change them in later life. That is why early childhood education and the early environment of children are so important.

I believe clinging to primitive beliefs has severely compromised the "moral" behavior necessary for our survival and future evolution. Eventually all life has either evolved or gone extinct. It's not clear we Sapiens are on a stable evolution path. We need to know that we control our future, to embrace it and take responsibility for that ownership.

I suggest we need a loud call for a self-examination not only of how we have structured our own cognitive map, but what it contains. Why is it put together that way, what is the environment in which it was constructed? Can it be changed based on new input? This process is not restricted to science. It is not what most understand as part of scientific thinking. I think often of the words of Percy Bridgman when asked to define science. *"Science is nothing more than doing your damnedest with your mind, no holds barred."* This can be adopted for everything in life.

Thomas Huxley, ~ 1850, put it well; *"The deepest sin against the human mind is to believe things without evidence."* That's why modern science was "invented" in the 17th century. Perhaps it was also why miracles were invented early in the historical time-line of humans. Many folks believe, live, and act in a cause and effect world and with the belief that any observable must have a cause. It's difficult to live with uncertainties, with things not yet known. It is very difficult to believe in a universe that came from nothing; cause and effect says there must be a causer, a God. Humans have made up a number of different gods. An authority is reassuring for many. Right now we have a Messiah-like politician who promises to make your life great again if voters will just put their trust in him. He has a secret plan, secret knowledge of what is needed. It sounds like that "good old time religion" shtick again. We also have a number of terrorist groups hiding in plain sight using the name of an established religion. For example, terrorist groups do not deserve to use the name Islam and to be called Radical Islamists.

Where are *Homo sapiens* going? We realize that we are the only humans left now on earth. Questions about our futures are many. How will we evolve? Will we evolve, remain the same, or go extinct? What path will we take?

I recall the response to the question on the probability of existence of life in the Universe by Enrico Fermi, today known as the Fermi Question, *"Where are They?"* This recognizes we live on a second generation sun system and there are many second generation sun systems in our universe with earth-like planets, some systems much older than earth, some planets where life, as we know it, has had more time to develop. These civilizations could have more advanced technology than we do. So why have they not contacted us? The answer to the Fermi Question may be that we are indeed alone; it may also be that any intelligent life that evolved went extinct a long time ago on all but our younger earth.

It seems doubtful that human cloning will happen very soon, if at all. What is much more likely is the development of reliable, low risk procedures to achieve gene expression (or non-expression) which will produce the desired change wanted in humans. Will we Sapiens then become wise gods? Recent reports out of China (Nature, April 2015) report limited success editing the genome of human embryos. Theyn discuss ethical concerns.

In a number of ways, we are already engaged in structuring environments. Much emphasis and study has been on a healthy environment during the nine-month long development in the mother's womb.

Changes in that small part of the earth that humans now can exist are troublesome and that is slowly getting increasing attention. It is not clear what environment and individual experiences are necessary so that individual truths include man-made effects of climate change and move to at least acknowledging the situation. Judging by the platforms of the two political parties in the US, we may be at most 50% there. It is not sufficient to just say, "the temperature goes up and the temperature goes down" and use this idea to dismiss the science behind the very strong story that predicts the most probable cause of global warming. Everybody may realize that this robust story predicts very unpleasant effects on the inhabitants of the earth. But, this should not be a reason to not consider the story's predictions on our future. To do so is beyond foolish.

Many societies and parents try to provide the "best" education, best health care programs, best living environment, and the best choice of a mate for their children. Providing quality environments for pre-school children is critically important. Unfortunately, as the world is always directed by adults, providing for those environments for all is often overlooked.

All Sapiens have basically the same DNA. There are no advanced beings in sight to interbreed with and to give us an evolutionary boost. So, outside of Aliens visiting, we cannot change by that route. It looks to me that what is left is to rely on technical schemes like gene splicing or future advances to control and change both DNA and gene expression. Only then will I believe in "intelligent design". Not in reference to a creator called God but in this case to wise humans. That is, if such a thing as a wise human exists.

Speaking of one's tribe or group as being exceptional is the first step to get comfortable with the idea of bio-engineering better humans, or in the extreme just killing off the "others". As we know the idea of better Sapiens has been around for a long time and some programs have been initiated. All have so far ended in failure.

I believe there is very little chance of "fixing" Sapiens so that they become true humans, or of developing a permanent engineering fix imposed on the environment. What we are likely to see in the near future is bio-technology advances that will allow a healthier life for a privileged group of Sapiens.

But, if a "good" change came to pass; if we are allowed to evolve to true humans, will we evolve globally into a very elite class and "the others"? Will the others slowly go extinct and the elite move toward a Star Trek future where money, resources, and religions are obsolete; where empathy is not a quality to be dismissed? John Lennon's beautiful song Imagine comes to mind. My money is certainly not on this future, but except for the others slowly going extinct part, my hopes are.

As E.O. Wilson points out in his new book, Half Earth, 2016, for almost any goals describing the future one can dream up and call ideal, not only most Sapiens but also your dog or cat would also agree with your goals. We have already made plans to colonize the planet Mars and there are a number of schemes to capture more energy from the sun, beam it to earth and store it. Traveling at 1/10 the speed of light, to visit and colonize earth-like planets in our galaxy takes about the same time than we humans have existed. So making such dreams reality is possible, but at present there is no evidence we will start down that path. As Wilson writes: *"Most of the time we behave like a troop of apes quarreling over a fruit tree."* Even if technological advances show the road can be traveled successfully, and we succeed in eliminating any aliens along the way to "colonizing that world" (recall that we Sapiens are practiced at that), we must first address the questions why, where are we going?

To again quote Yogi Berra: *"It's tough to make predictions, especially about the future."* It is easy to be pessimistic. There may actually be too little time left and more and more people are looking forward to the next life. Some are just not interested in issues like the possibility of human existence or extinction. There are people who claim their God tells them to kill others, people who claim their religion alone tells truth. So far, most organized religions get a grade of F in attempts to have people realize that they are responsible for their own actions; this is, I believe, a first step before we can seriously address the question where are we going. Few if any religions do this. Some do not deserve a grade of F, but it is the lowest grade I know to give. I am reminded of the line in the recent Batman Begins movie: *"it is not what you say, it is what you do that defines you."* It is for this reason that I greatly admire Pope Francis, the present Catholic Pope. He is trying to change what religion really means and to have people be responsible for their actions and to respect others. This sounds easy but he has a tough time working with the establishment of the Catholic Church. All organized religions need to do much better.

Present day science supports the stories presented here that are far more than ideas. I believe that the stories of the creation of our Universe, the creation of life, and the rise of Sapien consciousness are pretty strong stories, way more than just ideas. The path from ideas to stories, to individual truths, to facts is a complicated one. It is driven by the personal environment and consequences experienced by individuals.

Science and the individual search for truth are not anti-religious, but I am not sure religion is not anti-science. Most organized religions have their Truth and spend little time helping their followers to understand their individual truth.

I think often of a comment from my daughter Anne, 10 years old at the time; *"Dad, I used to worry about what happens to me when I die, but I don't anymore." I asked why. "Well, I figured it will be just like before I was born." Even that did not prepare me for her recent email question. "Dad, Quincy (her 10-year-old son) asked me the other day…Mom, what is the meaning of life? I thought I would ask you since you have lived longer than me and might know the answer…So, can you reply to Quincy; you have lived longer than me and may know the answer."* I found that question more difficult to answer than any question I have been asked.

For me, the answer certainly is not found in present day organized religions, or belief in an afterlife. I recall the words of Kathy's Dad, Tony, talking about a visit from a priest a few months before he died, "*It's not nice to fool people that way.*"

My life continues to be a quest to understand. Life offers a delightful opportunity to learn, to think, to interact with others, to be a leader along the future path of Sapiens. Enjoy it.

Finally, I share a memory of my first white-water raft trip. Mark had just pulled his shoulder and was unable to take us through the next class 4 rapid on the Salmon Middle Fork trip. I needed to do it. I went with him to scout the rapids and listened to his advice on reading the water, how to position the raft just before reaching the white water, and comments like to make sure the raft did not get pinned to a rock. I think I absorbed about 20 percent of what I was told. Mark's final words were "*don't screw up.*" All Sapiens and all religions please heed this advice. Remember the old Chemistry 101 T-Shirt label, *"If you are not part of the solution, you will be part of the precipitate."*

References

Web Resources:

Sean Carroll, *blog,* 2013 how-quantum-field-theory-becomes-effective
http://www.preposterousuniverse.com/blog/2013/06/20/how-quantum-field-theory-becomes-effective/

L.A. Hug et. al., *A new View of the Tree of Life*, Nature Microbiology, April 11, 2016,
http://www.nature.com/articles/nmicrobiol201648

New Tree of Life Doesn't Look as You'd Imagine, Earth and Sky, April 16, 2016
http://earthsky.org/earth/new-tree-of-life-doesnt-look-as-youd-imagine?utm_source=EarthSky+News&utm_campaign=40d842fdc4-EarthSky_News&utm_medium=email&utm_term=0_c643945d79-40d842fdc4-394592065

The Genographic Project, 2016
https://genographic.nationalgeographic.com/science-behind/

A Shocking Find in a Neanderthal Cave in Southern France, The Atlantic, May, 2016, http://www.theatlantic.com/science/archive/2016/05/the-astonishing-age-of-a-neanderthal-cave-construction-site/484070/

Lydia Pine, *Our Neanderthal Complex,* Nautilus, October 2014, http://nautil.us/issue/18/genius/our-neanderthal-complex

String Theory http://superstringtheory.com/

M. L. Glasser et al. *A multi-modal parcellation of human cerebral cortex,* Nature July 2016, http://www.nature.com/nature/journal/vaop/ncurrent/full/nature18933.html

The Human Connectome Project, https://www.humanconnectome.org/

2005 Pew Research, *Faith in Flux,* http://www.pewforum.org/2009/04/27/faith-in-flux/

2025 Pew Forum public becoming less religious, *http://www.pewforum.org/2015//11/03/i-s-public-becoming-less-religious/*

Nicolis Wade, lUCA, the ancestor of all living tings, http://www.nytimes.com/2016/07/26/science/last-universal-ancestor.html?em_pos=large&emc=edit_sc_20160725&nl=science-times&nlid=14715652&ref=headline&te=1

David Baker *the Hunter Captain*
http://wp.me/a4vwYm-3l April 2016

Emily Conover *LIGO's black holes may be dark matter,* Science News August 2016
https://www.sciencenews.org/article/ligo-black-holes-dark-matter?tgt=nr

Crisper Inspires New Tricks Science News August 22, 2016 https://www.sciencenews.org/article/crispr-inspires-new-tricks-edit-genes?utm_source=Society+for+Science+Newsletters&utm_campaign=2fcc65fb1e-editors_picks_week_of_082216_8_28_2016&utm_medium=email&utm_term=0_a4c415a67f-2fcc65fb1e-104498661

Have we been wrong about life's origin Earth Sky News August 28 2016 http://earthsky.org/earth/have-we-been-wrong-about-lifes-origin

Books:

J. Schweitzer and G. Notarbartolo-di-Sciara, *Beyond Cosmic Dice*, Jaquie Jordan Inc., April 2009

L. M. Krauss, *A Universe from Nothing*, Free Press, Simon and Schuster, Inc., January 2012

J.B. MacKinnon, *The Once and Future World*, Houghton Mifflin Harcourt, 2013

Richard Dawkins, *The God Delusion*, a Mariner Book Houghton Mifflin, 2006

Max Tegmark, *Our Mathematical Universe,* Vintage Books, 2014

Edward Frenkel, *Love and Math*, Basic Books, 2013.

Sy Montgomery, *The Soul of an Octopus*, Atria Books, May 2015

Edward O. Wilson, *Half Earth,* Liveright Publishing Corp.,2016

Edward O. Wilson, *The Meaning of Human Existence,* Liveright Publishing Corp, 2014

Erwin Schrodenger, *What is Life*, Cambridge University Press, 1967

Yuval Noah Harari, *Sapiens*, Harper Collins, 2015

David Wootton, *The Invention of Science*, Harper Collins, 2015

G. Kakari, *Soul Machine*, WW Norton and Co. Inc., 2015

Jo Marchant, C*ure, A journey into the Science of Mind over Body, C*rown Publishing Group, 2016

Weston La Barre, *Ghost Dance, the Origins of Religions,* Delta, 1972

Sean Carrol, *The Big Picture,* Dutton, 2016

John McPhee, *Annals of the Former World,* Farrar, Straus, and Giroux, 1982

De Waal, *are We Smart Enough to Know how Smart Animals Are?,* WW Norton and Co. Inc., April 2016

J.M. Marzluff and Tony Angel, *In the Company of Crows and Ravens,* Yale University, 2005

Fawaz A. Gerges, *A History ISIS,* Princeton University Press, 2016

Bernd Heinrich, *Mind of the Raven.* Harper Collins Press, 1999

Other Resources:

Science News: this is a biweekly publication by the Society for Science and the Public and covers recent advances in science. It is also available on the web at no cost at https://www.sciencenews.org/

The Great Courses: These are really great courses presented by award winning Professors.

Exploring the Roots of Religion John Hale, Liberal Studies, University of Louisville, 2009

How Jesus Became God, Bart Ehrman, Department of Religious Studies, North Carolina, Chapel Hill, 2011

The History of the Bible; the Making of the new Testament Canon, Bart Ehrman, Department of Religious Studies, North Carolina, Chapel Hill, 2005

Understanding the Brain, Jeanette Norden, Vanderbilt University School of Medicine, 2007

The Long Shadow of the Ancient Greek World, Jan Worthington, History, University of Missouri-Columbia, 2009

History of the Ancient World: A Global Perspective, Gregory Aldrete, Humanities Studies and History, University of Wisconsin-Green Bay, 2011

The Rise of Humans, John Hawks, Anthropology, University of Wisconsin-Madison, 2011

The Story of Human Language, John McWhorter, Manhattan Institute, 2004

Cosmology: The History and Nature of Our Universe, Mark Whittle, Astronomy, University of Virginia, 2008

The Higgs Boson and Beyond, Sean Carroll, Physics, California Institute of Technology, 2015

Mind-Body Medicine: The New Science of Optimal Health, Jason Satterfield, University of California San Francisco, 2013

Understanding the Universe, Alex Filippenko, University of California, Berkeley, 2007

www.ingramcontent.com/pod-product-compliance
Lightning Source LLC
Chambersburg PA
CBHW070319190526
45169CB00005B/1667